Addressing the Blues: Handling Depression

Addressing the Blues: Handling Depression

Shivsanjeevi Sripathi

Delve PUBLISHING

www.delvepublishing.com

Addressing the Blues: Handling Depression

Shivsanjeevi Sripathi

Delve Publishing
224 Shoreacres Road
Burlington, ON L7L 2H2
Canada
www.delvepublishing.com
Email: orders@arclereducation.com

ISBN: 978-1-77469-043-7 (Hardcover)

ABOUT THE AUTHOR

ShivSanjeevi Sripathi completed his Masters in Biotechnology from Mumbai University in 2008. He was awarded for academic excellence in both his Bachelors and Masters for securing second rank in Mumbai University in 2006 and first rank in his college: Kishinchand chellaram College. For his Masters he secured first rank in his college KET's V.G.Vaze College. He qualified CSIR and NET and and TOEFL in September 2008. He then worked on a stem cell project at the Specialized Centre for Cell Based Therapy (SCCT),KEM Hospital at Mumbai on a project entitled, : Isolation & detection of stem cells from Human Umbilical cord/ amniotic membrane" following which he worked at Junior Research Fellow at Microbiology & Cell Biology Department, Indian Institute of Science, Bangalore on cloning of cell wall genes and transcription factors in E.coli & M.smegmatis. As a writer, he has authored and co-authored 35 books on various aspects of biology such as bionics, molecular wires, cloning, hypertension, the epidemics of the 21st century, handling depression, camouflage, hygiene, immunology and many more with international publishers. He loves to read and share on interesting aspects of life sciences in books. In his free time he loves to travel and explore and give talks on spirituality, ancient customs and traditions.

TABLE OF CONTENTS

ACKNOWLEDGEMENT

Depression has now been termed as “a common illness”. This condition is seen right from adolescents to the elderly. This book is a dedication to all who have battled this condition and are still battling it. While external treatment approaches play a vital role, our inner attitude and confidence can enable us to tide over the “blues”.

This book is a dedication to *parents* Mrs. & Mr. Sripathi/ Mrs. Pandey who have stood by us through thick and thin and are the biggest sources of unconditional love , and *teachers* like Mrs. Rajalakshmi Ratan, Mrs. Monali Shriyan, and Mrs. and Mr. V. Shankar Srinivasan who have been “lighthouses” during the blues.

Thank you, Mr. Sushant Singh Rajput for making me believe in dreams. I dedicate this book and a quote of his to him and all those going through tough times: “*We should never forget the inevitable, as we will lose everything eventually. So, why fret over any kind of security? The idea is to just fly and experience it all while it lasts*”

—Sushant Singh Rajput.

This book is also a dedication to Dr. Prerna Pandey who is a rock of support and Mrs. Madhumati Subramaniam who has been an inspiration.

Special mention to Ms. Pooja, Gayatri, Garima, Anant!

Grateful to the Divine and Masters!

इयंवसृिष्टर्यितआबभूवयदविादधेयदविान।

योअस्याध्यक्षःपरमेव्योमन्त्सोअङ्गवेदयदविानवेद

- Whence all creation had its origin,the creator, whether he fashioned it or whether she/he did not,the creator, who surveys it all from highest heaven, she/he knows — or maybe even she/he does not know-Nāsadīya Sūkta, Rig Veda.

Dedicated to each corona virus warrior, worker and survivor: Thank you!

PREFACE

"*If you know someone who's depressed, please resolve never to ask them why. Depression isn't a straightforward response to a bad situation; depression just is, like the weather. Try to understand the blackness, lethargy, hopelessness, and loneliness they're going through. Be there for them when they come through the other side. It's hard to be a friend to someone who's depressed, but it is one of the kindest, noblest, and best things you will ever do.*"

— Stephen Fry

According to the WHO Fact Sheet (January 2020), the numbers of people globally depressed are more than 264 million people! The sheet goes on to describe that depression is "a common illness". In the case of adolescents, the prevalence of depression varies from 1.1 to 14.6% (Beirão et al, 2020, Middle East Curr Psychiatry). Depression is complex: take for instance, research in Scotland showed that 19% of least advantaged groups was depressed while 10% of those who were the most advantaged (Barnett et al, Lancet, 2012).

"*Whenever you read a cancer booklet or website or whatever, they always list depression among the side effects of cancer. But, in fact, depression is not a side effect of cancer. Depression is a side effect of dying.*"

— John Green, The Fault in Our Stars

If we take into account chronic diseases such as cancer, the prevalence of depression and anxiety among people with cancer according to evaluations are reported as 5% to 16% in outpatients, 4% to 14% in inpatients, 4% to 11% in mixed outpatient and inpatient samples and 7% to 49% in palliative care (Walker et al, *The Annals of Oncology*,2013).

Looking at the causes, contrary to what is thought that depression happens after a "sad event" or "loss", a combination of genetic, biological, environmental, and psychological factors has been identified by on-going studies. Given the multiple factors associated with this complex condition, further studies shall help comprehend the exact causes.

In the recent COVID-19 epidemic, studies have shown that the prevalence of depressive disorder was around four times (24.3%) than what was reported in the second quarter of 2019 (6.5%) showing how depression has affected our lives due to a pandemic (Czeisler MÉ, Lane RI, Petrosky E, et al. Mental Health, Substance Use, and Suicidal Ideation During the COVID-19 Pandemic — United States, June 24–30, 2020. MMWR Morb Mortal Wkly Rep 2020;69:1049–1057).

This book comes at a time when the world is on the path to recovering from the aftermath of a viral epidemic: healing at physical, emotional, financial, social, and mental aspects.

The book presents factors identified as causes of depression backed by research such as the COVID-19 epidemic, social media usage, and social displacement. The social and economic aspects have also been presented to further comprehend how this condition affects several aspects of living. A section has also been presented on the trends of depression in adolescents and how to address the condition.

While little focus has been given to antidepressants used for treating depression given the vast plethora of matter available and the requirement of personal diagnoses followed by treatment recommendations by a qualified expert, other approaches where one can seek relief have been discussed. These encompass yoga to meditation to mantras to guidelines in the usage of antidepressants to clinical guidelines to handle depression: all backed by solid scientific research.

"*The world is indeed full of peril, and in it there are many dark places; but still there is much that is fair, and though in all lands love is now mingled with grief, it grows perhaps the greater.*"

— J.R.R. Tolkien, The Fellowship of the Ring

We present to you new insights to handling depression: pointers that can be kept in mind while discussing with a healthcare provider during the design of the treatment modules for one's self or relative/friend.

An ancient quote from the Upanishads of India says "तमसोमाज्योतिर्गमय" or lead us from darkness into light". Let us delve into the concepts and approaches to cast light on the darkness and handle the blues!

Disclaimer: Though all the content presented in this book is based on scientific research articles, the approaches or molecules named should not be taken without medical consultation.

Cover image source: Photo by Samuel Clara on Unsplash:

(*Unsplash photos are made to be used freely. Our license reflects that.*

a. All photos can be downloaded and used for free

b. Commercial and non-commercial purposes

c. *No permission needed (though attribution is appreciated!*)

CHAPTER 1

DEPRESSION: LET US START WITH AN OVERVIEW

CONTENTS

Mental health and disorders have become the most discussed topics over the recent decade, although over several years, despite several research and studies conducted, it was considered a taboo for a very long time. In our day-to-day conversations with our friends, colleagues, family, etc. it has been observed that people often casually label others who function a bit differently or not comfortable in certain situations. To state a few examples, being neat, orderly, does not make one an OCD.

Similarly, a person is often labeled as someone suffering from depression because he may have been unhappy in the recent past. But it need not be necessary that the person is suffering from depression. It could be that they are taking time to recover from a certain event that has rendered them disoriented. Before we casually label someone as a depression patient, let's understand what is depression is.

1.1 DEFINITION OF DEPRESSION

According to the Oxford English Dictionary, Depression is defined as "a mental condition characterized by severe feelings of hopelessness and inadequacy, typically accompanied by a lack of energy and interest in life." So next time we think someone is suffering from depression, we need to understand that they are just feeling sad.

Is being depressed and suffering depression different from each other? Yes, pretty much. Being depressed often has to do with the usual sadness that a person goes through. It can be due to numerous reasons, say a loss of a loved one, losing a job, or having a rough day. This type of sadness doesn't last long. However, Depression is a mental condition and is chronic. Commonly, people cannot tell the difference between a depressive episode and an actual disorder. Let's take a detailed look into the difference between both.

As per the classification systems used by the World Health Organization (WHO), ICD-10(International Classification of Diseases, 10th edition) defines **Depressive episodes** as where the patient undergoes a reduction in activity, decrease in energy, and lowering of mood. Enjoying capacity, concentration is reduced, and there is tiredness even with minimal activity. Sleep is reduced, and appetite is diminished. The lowered mood varies from day-to-day and is unresponsive to various circumstances and is sometimes accompanied by somatic symptoms, such as loss of interest, oversleeping, unable to feel pleasure or happiness from the one usually does, loss of appetite, and loss of libido.

A recurrent depressive disorder is characterized as repeated episodes of depression as defined for depressive episodes. Without any history of independent episodes of mood elevation, and increased energy also termed as mania. There may be some mild mood elevations for a brief period and then hyper-activity after the depressive episode which is precipitated with anti-depressant treatment. However, more study is being conducted to understand when a depressive episode turns into a recurrent depressive disorder. Study shows an episode may take place anytime between childhood and old age. However, an episode becoming a disorder solely relies on various factors of an individual, their surroundings, lifestyle, etc.

There are various shades of depression. The symptoms of a depressive episode include loss of interest, depressed and increased fatigability. Based on the severity of the symptoms, a depressive episode can be classified as mild, moderate, or severe. People suffering from mild depressive episodes may face a little difficulty in continuing their daily activities but not to such an extent that hampers their daily living. However, during a severe depressive episode, an individual won't be able to carry on with his daily activities and hampers progress to a great extent.

Depression and physical health are very well connected. For example, a person suffering from fatal heart disease can trigger depression. And a person's physical health can also deteriorate when they are suffering from depression.

Various pieces of evidence suggest the need for further classification of depression based on clinical commonalities, neurophysiopathology, etc. Certain classifications under bipolar conditions include rapid mood cycling, bipolar depression, dysphoric mania, etc. Other clusters would something more severe such as OCD(Obsessive Compulsive Disorder). And the clusters would include general anxiety, panic attacks, PTSD(Post Traumatic Stress Disorder), etc.

1.2 ORIGIN

Taking a look back at history, numerous personalities have suffered depression. Right from Ramayana's protagonist Lord Rama to everyone's beloved inventor of the modern world, Nikola Tesla. However, their causes of depression are not mentioned in detail but it mostly has been situational and was probably just a recurrent depressive episode which they later overcame. Let's talk about how it begins.

In early adulthood, depression is said to begin with likely recurrences. And acknowledging the biological aspects of the body, depression is said to affect women more than men. And also unemployed people are at higher risk. An episode may be classified by indifference, sadness, apathy, or irritability. It is usually associated with changes in bodily functions such as sleep patterns, loss of appetite, impaired concentration, etc, which is accompanied by feelings of shame or guilt and suicidal thoughts in most cases. A minuscule portion of patients will experience psychotic symptoms.

Unfortunately, the nature of depression is such that people don't realize that they are affected by it and therefore, won't reach out for professional help. In long-term conditions, the appearance of depression highly impacts the quality of care provided by the patient and received from others.

Depression in those who are physically unwell is quite common and is highly responsible for the cause of morbidity. The risk of major depressive disorder(MDD) is highly likely in those physically ill. As we go further, we will come across various surveys and statistical analyses of depression done by various research centers. Many reviews and surveys show that the prevalence of depression amongst various patients with different ailments such as diabetes, cancer, Parkinson's disease is also high.

When treatment for an ailment is not administered properly it heightens the risk of depression. Ideally, it's not the physical ailments that cause mortality among the patients but the depression that comes with it which responsible for deaths among the patients. Depression can also trigger various mental disorders which can hamper ones' living and the mental recovery becomes utterly steep.

1.3 DEPRESSION AMONG ADOLESCENTS AND YOUNG PEOPLE

Depression is common among many adolescents and children, affecting many before the age of 18. It is stated that every 10 in 100 adolescents are affected by depression in the USA. These results are in congruence with studies that show the mood disorders among children and adolescents are more common than believed. Depression among adolescents affects their social life, family relations, and their academic performance which potentially pose long term consequences. Adolescents suffering from depression are at risk of psychosocial impairment, alcohol abuse, and recurrent depressions as they grow up.

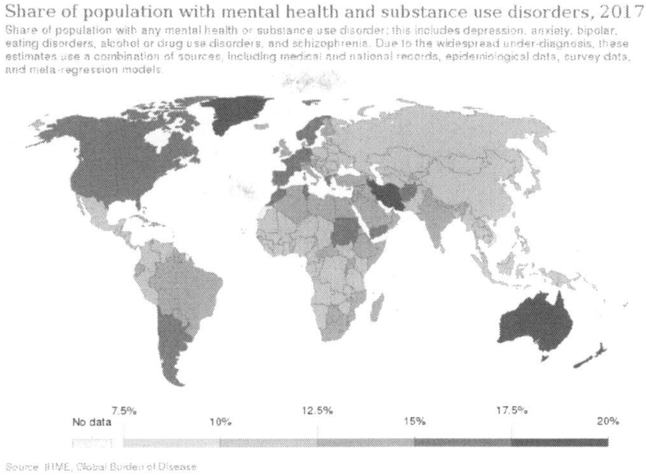

Figure 1.1: Looking at depression.

Image source: https://ccsearch-dev.creativecommons.org/photos/3ecefce2-43ae-4921-95d8-2676aec8aad2 (Commercial use allowed)

Among adolescents, depression often is associated or caused because of poor academic performance, social dysfunction, teenage pregnancy, and substance abuse. It is also relatively connected to attempted suicide or loss of a close one. This makes depression not only a problem to one who's suffering from it but also to the family and close circles of the patient.

Figure 1.2: The stress.

Image source: https://ccsearch-dev.creativecommons.org/photos/38f7c1a4-bc77-485a-82e2-9f982e4f4d54 (Commercial use allowed)

Depression among youngsters is expressed in different ways and even affects them in different ways. In most cases, it leads to substance abuse, concurrent psychiatric problems. It can also manifest into many behavioral disorders such as misconduct, verbal aggression, irritability, etc.

The impact of depression differs among children, adolescents, and young adults. Among children, it may manifest as a depressive episode creating a short-term impact and can be overcome with numerous activities. Among adolescents, it may last as long as one year but if not treated properly it may have long-term consequences.

Among adults, the impact is pretty severe. It depends on the situations, people, and societal pressures they succumb to. Often it adds to low self-esteem, lack of motivation, anger, low-performance, suicidal thoughts, and tendencies. Depression also leads to hopelessness and overeating although the reason for variations in depressive symptoms is unknown.

The expenditure that is incurred to treat depression among adolescents is higher as compared to children with other issues or no issue at all. It is found that children with depression use emergency and inpatient services more than other children.

1.4 DEPRESSION AMONG ELDERLY PEOPLE

Depression affects people are all age groups. But the most affected are adolescents and elderly people. There is a growing concern for the prevalence of depression among the elderly, particularly in Western Europe. Depression is quite common in people with weak physical health. In a normal medical setting, this is due to the result of poor outcomes, longer stays, and compromised care.

These outcomes make depression even trickier to treat and more expensive in terms of healthcare costs especially among the elderly patients who are subject to poor physical health. Thus, the burden is higher among the age group of 55-59 who are subject to major depressive disorder.

Behavioral and somatic symptoms associated with depression are ideally so intense that it covers the psychological symptoms which may lead to a point where an individual might be suffering from clinical depression without sadness. The simultaneous occurrence of several chronic conditions and symptoms further complicates the diagnosis and the treatment. In cases like these, the symptoms may show the psychological stress of enduring or coping with a disease, which may have been caused by the disease process or due to medications provided to treat the disease or it may even manifest during the medical diagnosis. The elderly are more likely to be under medications for the treatment of diseases that comes with age. Women who undergo menopause are more likely to suffer from depression. Many medications or prescribed drugs can be a potential cause of depression as many may hamper hormonal processes or cause chemical imbalances in their body.

Figure 1.3: Old age and depression.

Image source: https://ccsearch-dev.creativecommons.org/photos/7be364c1-3233-40b1-995e-9bc6cac08c77 (Commercial use allowed)

Depression is diagnosed in elderly patients possibly because of years of neglect and delay. A study in the UK suggests that approximately 15% of the elderly patients whose depressive illness goes unnoticed are likely to commit suicide. The same study also stated that most elderly patients had visited physicians one or two months prior to the event where they commit suicide but it is because the symptoms were not recognized or traced and treatment provided was not adequate.

Depression is a disorder that works at a very subtle level and cannot be traced easily. However, earlier symptoms can be detected if the concerned patient opens up about his condition and seeks professional help.

(Cesar, Julisca, April 2013) (Chavoushi, Faraz, April 2013)

1.5 LONELINESS AND DEPRESSION

Loneliness is a common experience or a phase that an individual goes through when they are having a depressive episode. When they are subject to a break-up, a loss of a loved one or have undergone something tragic, they prefer to left alone to recuperate and heal on their own. It is not wrong to be alone for some time as it is a natural tendency to process a particular and overcome it in a period of time. However, an episode goes beyond and takes longer it may manifest as a depressive disorder.

Figure 1.4: The heartbreak and grief.

Image source: https://ccsearch-dev.creativecommons.org/photos/4148b533-c0de-4ccc-8395-971d87038134 (commercial use allowed)

However, loneliness can manifest through mediums as well. It can be termed to that aspect as well where an individual is not satisfied with human relations and does not like to mingle socially. According to various research and studies, loneliness has a significant amount of implications associated with various mental and physical health issues such as depression, cardiovascular problems, insomnia, Alzheimer's disease, and even depression.

To understand the impact of loneliness on an individual and mortality related to it, the gender analysis must be done carefully. It is given that women live longer than men and feelings of loneliness are more prevalent in women than men. Men and women network differently, for example, the events women want to go can be different from men, be it kitty parties, pubs, concerts, movies, etc. The articles even suggest that depression is more prevalent in women than men. However, it claims can be challenged as socially men are not encouraged to be emotional than that of women.

There is a lack of proper gender-based analysis on how loneliness affects the mental health and livelihood of an individual. Each individual is capable of feeling lonely and often those are for very good reasons. But however, if it is left unattended and if the individual refuses to reach out for help, it may lead to dire consequences that may affect both physical and mental health alike.

There is a significant difference between alone and feeling lonely. A person maybe well contented when they are alone and are happy in their own company. By happy, we are talking about healthy living both physically and mentally. However, there are several cases of an individual where they are surrounded by people but they don't feel belong in that social group. This leads to feeling lonely and often leads to social anxiety and depression.

The impact of loneliness can be analyzed through various reviews and literature. However, loneliness depends on how an individual takes it and deals with it. It also deals with what situations or circumstances that a person is going through. Although in some cases, loneliness can also be associated with disorders and psychoses, an increase in depressive symptoms. There is a fine link between depressive symptoms and loneliness. Many articles suggest that the increase in depressive symptoms due to prolonged loneliness but depressive symptoms do not increase loneliness and may push an individual to either actually reaches out for help or may take up extreme measures such as self-harm, suicide, etc.

One of the biggest damages that loneliness can do is dementia and cognitive decline. To feel lonely at an old age say in the late 70s can lead to lifetime cognitive change. This indicates a lower IQ and adjusting to the IQ of an 11-year-old. The studies also acknowledge that cognitive impairments due to old age can also hamper social interactions, deliberate social withdrawal, which significantly leads to loneliness. There are circumstances where loneliness leads to Alzheimer's and vice-versa.

However, depression symptoms have the least impact on Alzheimer'sdisease risk. However, loneliness has a significantly higher impact and may lead to several complications and other mental disorders.

It cannot be entirely blamed on loneliness for the cause of the mental disorder as it also depends on the lifestyle of the individual who is living a secluded life. Various factors such as physical health regime, hobbies, and to great extent the ambiance of the environment that individual is subject to.

As mentioned earlier, loneliness need not mean being away from social interaction. It can also be situations where the person may be socially active but they may not feel they are with the right tribe. Here, the individual actually yearns for a true company and thus, leads to depressive episodes. If this was not properly treated, this may manifest into irritability, aggression or even anger.

As we dwell further, we will take a look into how depression affects not only our physical health but also injures the relation with the social circle which makes an individual be alone that further damages their well-being(Hawkley, Louise C, Dec 2013; Cacioppo, John T, Dec 2013).

1.6 VARIOUS MANIFESTATIONS OF DEPRESSION AND CAUSES

Every condition should have a reason for cause and the purpose of research is analysing what exactly causes a particular condition. There are various manifestations and causes of depression as mentioned in DSM-5, a manual where assessments and diagnosis of mental disorders are recorded. It does not include guidelines or treatment for any disorder specified. Below are some of the manifestations of depressive disorders.

1.6.1 Persistent Depressive Disorder

This is also known as Dysthymia and it is defined as a chronic mental disorder as per the DSM-5. It is characterized as a depressive mood for a few or more days but not longer than 2 years. Among children and adolescents, it may last up to 1 year, and where they show signs of irritability. If it persists way beyond 2 years it can be diagnosed as Persistent Depressive Disorder. The symptoms in this disorder vary between, insomnia or hypersomnia, over-eating or loss of appetite, low-energy or hyperactive, feelings of absolutely hopeless and have difficulty in making decisions.

1.6.2 Medication-Induced Depressive Disorder

This is type of disorder is caused due to presence of any medication or drug prescribed. Some medications may generate symptoms, and it is important to understand whether the symptoms were actually caused due to medications or whether the symptom was pre-existing even before the medication was prescribed. Some patients are prescribed sleeping pills which have a deeper impact on their physiological well-being as a result affects their mental well-being too.

1.6.3 Disorder Caused by a Medical Condition

In many cases, a depressive order is caused due to the medical condition an individual is diagnosed with. When an individual is diagnosed with a fatal disease such as cancer, heart ailments, etc. it mentally affects the individual which makes them withdraw from all social activities leading to prompt loneliness. The constant fear of death or having the day's numbers plunges the patient into a depressive mood which turns into a disorder after an episode gets prolonged. Sometimes, the disorder may get amplified due to medications prescribed to tackle the said medical condition.

1.6.4 Premenstrual Dysphoric Disorder

As DSM-5 suggests the majority of the menstrual cycles, there are at least five symptoms that will be present during the last week before the start of a menstrual cycle, and the individual in question begins to feel better after a few days where all the symptoms disappear the week after menstruation. The characteristics of the disorder manifest as anger, irritability, anxiety, conflict in interpersonal relations, etc. These symptoms must be prevalent mostly during the menstrual cycles and affect social functioning.

1.6.5 Single Depressive Episode

The classification of this depressive episode can be varied on the basis of severity. It can be mild, moderate or severe without any psychotic symptoms. Some of the common characteristics of this mood are low energy, reduction in mood, and reduction in daily activity. There is also a loss of libido, reduced concentration, fatigue, and tiredness even after doing an activity that requires minimum effort. Changes take place in appetite, sleep patterns, self-esteem, and even a drop in self-confidence. In mildest forms, these symptoms may last up to 2-3 symptoms but manage to carry on with their daily lives. In moderate forms, there may be 4 or more symptoms but will have difficulty continuing their daily activities. In the severe forms, several symptoms are visible and last for several days, the symptoms are dangerous and distressing which leads to feeling worthless, which may lead to suicidal tendencies. If accompanied with psychotic symptoms, the presence of delusions, hallucinations, retardations are detected which makes carrying out daily activities impossible. This potentially even poses threat to life either through suicide, dehydration or even starvation as there is a loss in appetite.

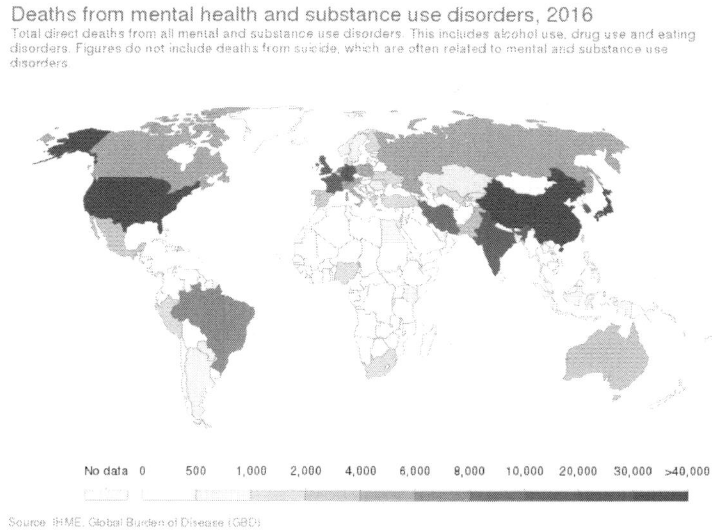

Figure 1.5: Depression and death.

Image source: https://ccsearch-dev.creativecommons.org/photos/3287e281-0060-432c-89bd-c73dad4bf563 (commercial use allowed)

1.6.6 Recurrent Depressive Disorder

When a depressive episode repeatedly takes place over a prolonged period of time, it slowly grows into Recurrent Depressive Disorder. In mild scenarios, there may be short-term (meaning brief) episodes of mood elevations and hypomania (over activity) after a depressive episode. Severe forms of disorder lead to manic depressive disorder, melancholia, endogenous depression, and vital depression. The first episode may happen anytime from childhood to old-age and may range from few weeks to few months.

1.6.7 Persistent Mood and Other Mood Disorder

Persistent mood disorders are classified as those moods or disorders which are persistent and are usually fluctuating in which the majority of the episodes are not severe enough to warrant as hypomanic or depressive episode of mild severity. If it becomes too severe then it may lead to cyclothymia or dysthymia.

In other mood disorders, it includes any disorder that does not satisfy the criteria of the above-mentioned disorders because they are not severe enough or may not last for a longer duration.

The above-mentioned manifestations and symptoms of depression vary as per the age and level it is developed to. However, DSM-5 has never distinguished or differentiated between child and adult depression. It also takes into that children and youngsters between the ages 7 to 18 express distress in different ways. It could manifest as recurrent irritability which may be expressed verbally or behaviorally. In major depressive disorder among children, the mood is irritable rather than depressed.

1.7 THEORIES OF DEPRESSION

There is no single theory that can explain depression. There are many variables involved on different levels. Therefore the 2 main causes that are taken as a reference for this section are biological factors and psychological theories. Under both factors, there are subsets that are responsible for the cause of depression:

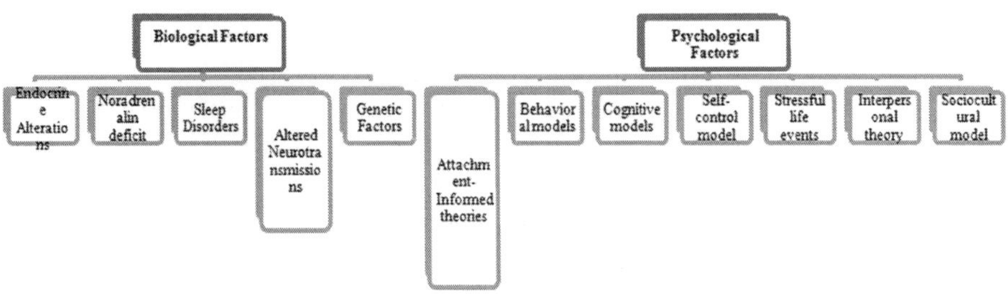

Figure 1.6: The scheme of depression.

The above flowchart shows the subdivisions under both biological factors and psychological factors.

1.7.1 Biological Factors

A mood disorder may not be explained by family history or stressful scenarios, it is likely that child or the adolescent is suffering from a neurological disease. Below are some of the deficits and factors that are biological in nature:

Endocrine alterations:

Many age-related changes and the presence of biological factors that pose a certain risk which include endocrine, immune or inflammatory, cardiovascular factors can make an individual vulnerable to depression. Various studies suggest that depression can be linked to endocrine alterations and that hormonal conditions or a chemical imbalance in the body can lead to depression.

Sleep disorders:

Various situations can lead to sleep deprivation or insomnia. Social deprivation, unemployment, or stressful life events can be some of the factors. Prolonged sleep disorder leads to deterioration of physical health while slowly builds into a depressive disorder. Studies show that insufficient sleep affects the hippocampus and makes an individual vulnerable to neurotoxic challenge.

Noradrenalin deficit:

Adrenaline based deficit or imbalances in the body are linked to depression. Serotonin is the key chemical that is linked to dopamine, norepinephrine and adrenaline which play an important role in life regulatory functions such as appetite, sleep, learning, memory, social behavior, etc. Serotonin contributes to neuroplasticity in the early years of life and dysfunctions in these systems contribute to the physiopathology of depression.

Genetic factors:

Many studies have emphasized the importance of genetics in depression. It is necessary to understand that genetic predisposition can trigger depression among children if they are exposed to stressful situations, anxiety or any stress-related disorder.

These are some of the theories related to the biological factors which may trigger depression. Biological factors, however, can be taken care if the environment is correct or the right treatment is given at the right time in order to avoid collateral damage in the future.

(Bernaras, Elena, March 2019)

1.7.2 Psychological Factors

This section mainly touches on the psychological aspects of how and why depression comes into existence. There are several theories and reasons for depression to take place in an individual. Although, it is clear there is no one single theory that can explain the origin of depression, let's take a look at possible theories for the origin of depression.

Stressful life-events:

Studies have found that the majority of the adults say 60-70% faces one or more stressful life event prior to having the onset of depression. In adolescents and teenagers, moderate amount stressful life event has been found that is linked to their depression. It can vary from academics, peer pressure, friends' circle, bullying, etc. Individuals who had gone through emotional abuse in their childhood are more likely to have levels of depressive symptoms. Even minor life events such as failing in academics during school, financial difficulties in family, betrayal, etc. can trigger depressive symptoms.

Socio-cultural models:

Cultural variables can also dictate and trigger depressive symptoms. There are two variables to this model, Acculturation and Enculturation. Acculturation touches upon structural changes (demographic, political, and economic) alongside people's change in mindset (psychological behavior). Observations hint that increases in suicides are due to economic scenarios (recession, inflation, unemployment).

Enculturation occurs when the previous generation induces or forces the traditional mind on to the younger generation. This creates a conflict or friction between the generations leading to fights, disagreements and even disowning of either of the party.

Interpersonal theory:

This theory sheds some light on attachment theories, which aim at identifying an individual's struggle with depression associated with interpersonal relationships and their functioning. It states the difficulties are closely linked with unresolved grief, transition of roles, disputes, and interpersonal deficits. Many researchers argue that pathological attachments during childhood can serious consequences for adults' to understand and sustain positive relationships.

Attachment theories:

This theory talks about humans' tendencies to establish strong and long-lasting ties with people create a considerable impact on their well-being. Consistency, nurturance, protectiveness, and responsiveness in a relationship make an individual dependant on a person. However, when an aspect of a relationship gets injured, say breaking of trust, betrayal or even death of the person with whom the attachment was maintained can trigger depressive symptoms.

Self-control model:

In this aspect, it is assumed that depression is due to mishandling or deficits in self-control, which broadly categorized into 3 sects and the consequences curtailed to it, namely self-monitoring, self-evaluation, and self-administration. In self-monitoring, an individual only attends to negative events and tend to recognize only immediate or short-term consequences. In self-evaluation, an individual sets unreachable or impractical criteria of their idea of success in life. In self-administration depends on self-evaluation that if it turns out to be negative, the individual engages more in self-punishment and very less in self-reinforcement.

Behavioral models:

The explanation revolves around a lack of positive reinforcement, avoidance of behavior in excess quantity or loss of efficiency in the positive reinforcement and depression linked to it. Some of the common scenarios are when a child with depression gets attention for their behavior such as crying, complaints, etc. When these depressive symptoms increase, the friends or people on whom the child relied on for support begin to avoid or be averse of the child, therefore aggravating the already existing depression in the child. Lack of supportive friends' circle or lower parental support are some of the signs of lack of positive reinforcement which are some primary causes of depression among children and adolescents, which also affects their future relations as it also triggers trust issues.

To summarize, according to different theories as mentioned above depression can be caused due to multiple reasons such as biological reasons, sociocultural changes, negative attributions made by individuals about themselves, negative or toxic interpersonal relations, lack of positive reinforcement, negative self-evaluation resulting in self-punishment. Therefore, depression cannot be attributed to one single point of origin as it can stem from various factors(Jaureguizar, Joana, March 2019).

1.8 DEPRESSION PREVENTION PROGRAMS IN THE SCHOOL ENVIRONMENT

Therapy has been advised by many who are diagnosed with depression. Among children and adolescents, cognitive behavioral programs are initiated by teachers with specialist training or psychologists. Ideally takes about 8 to 15 sessions depending on the severity of what the child is going through. There are only a handful of programs that are designed to prevent the symptoms of depression in younger children since most affected are adolescents and teenagers.

Some results are not particularly positive as said earlier, the reason for depression doesn't come from a single point hence it gets trickier to analyze the problem in order to find the solution. As the child cannot understand what is bothering them or who don't reach out easily it's nearly impossible to even trace if there is any depression.

1.9 CLINICAL TREATMENT FOR DEPRESSION: A SNAPSHOT

The World Health Organisation (WHO) says that prevention programs reduce the risk of depression and the suffering that comes with it. But it is yet to confirm what kind of programs and what contents are most effective. As the methods of treatment may vary from person to person, WHO says that cognitive-behavioral therapy is more effective. However, for moderate and severe depression can be treated with interpersonal psychotherapy, cognitive behavioral therapy and anti-depressant medication. However, the effectiveness of all methods of treatments can only be ascertained if there are thorough follow-ups, regular sessions, consistent analysis of the core issue, and honesty from both the party (patient and therapist). If there is a slight compromise from either party then the result may not be accurate and the treatment may not be effective.

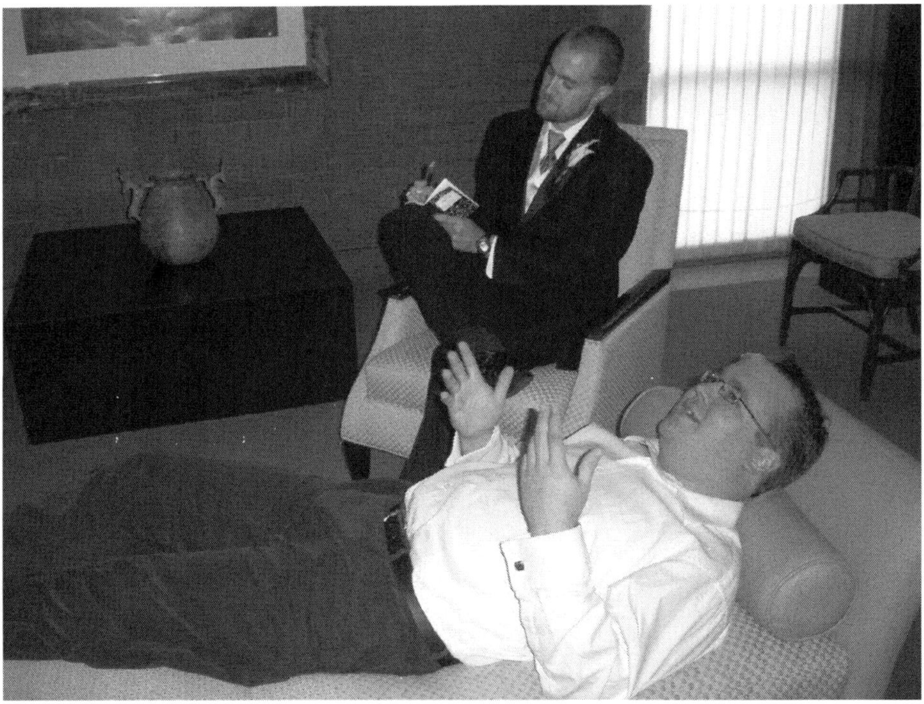

Figure 1.7: Talking depression away.

Image source: https://ccsearch-dev.creativecommons.org/photos/d9869272-e083-42e9-b6cb-543d3cd81725 (Commercial use allowed)

Some of the known treatments and that are prescribed to the patients:

- Self-management therapy
- Cognitive therapy
- Interpersonal therapy
- Behavior therapy

(Bernaras, Elena, March 2019; Jaureguizar, Joana, March 2019; Garaigordobil, Maite, March 2019)

1.10 REFERENCES

1. Elena Bernaras et al, March, 2019, Child Adolescent and Depression
2. Joana Jaureguizar et al, March 2019, Child Adolescent and Depression
3. MaiteGaraigordobil et al, March 2019, Child Adolescent and Depression
4. Julisca Cesar, et al April 2013, WHO Background paper 6.15, Depression
5. Faraz Chavoushi et al April 2013, WHO Background paper 6.15 Depression
6. Louise C. Hawkley, et al Dec 2013, Loneliness Matters
7. John T. Cacioppo, et al Dec, 2013, Loneliness Matters

CHAPTER 2

DIFFERENT INSIGHTS OF DEPRESSION

CONTENTS

This chapter explores insights into the many causes that lead to feeling depressed, as reported by research.

2.1 SOCIAL MEDIA: A TWO-WAY STREET

Whether it is while traveling or waiting in a queue, one can observe the trend of many people peering into their phones!

Social media are the platforms powered by the internet for visual and verbal communication. Given the social nature of humans and the necessity for social connections by many, various social media platforms such as Facebook, Messenger, Instagram, etc. have emerged as integral parts of many of our lives (Keles et al, 2020; Karim et al, 2020).

Figure 2.1 below shows the usage of social media platforms segregated by gender (data as of January 2020 (Karim et al, 2020: This is an open-access article distributed under the terms of the Creative Commons Attribution License, which permits unrestricted use, distribution, and reproduction in any medium, provided the original author and source are credited):

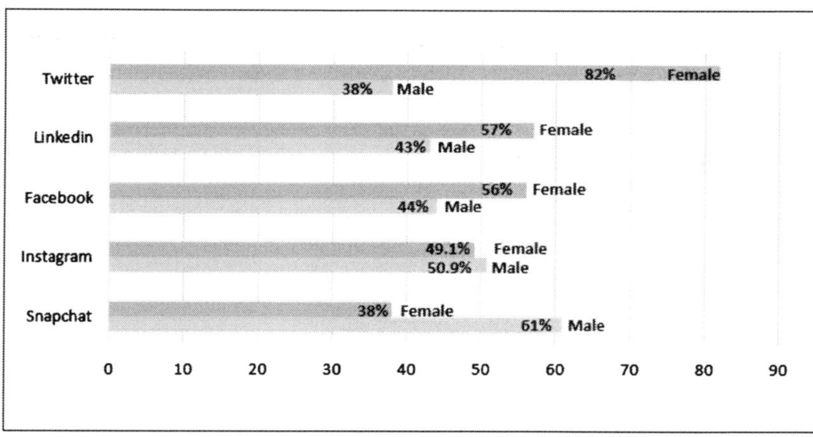

Figure 2.1: Above shows the usage of social media platforms segregated by gender (data as of January 2020 (Karim et al, 2020: This is an open-access article distributed under the terms of the Creative Commons Attribution License, which permits unrestricted use, distribution, and reproduction in any medium, provided the original author and source are credited).

While on one hand, the use of social media is considered enjoyable and offers benefits (Karim et al, 2020); there are always 2 sides to the coin. For instance, a 2017-published research article as part of the European School Survey Project on Alcohol and Other Drugs (ESPAD) gauged the prevalence of problematic social media usage. Following the social media usage of 5,961 adolescents from Hungary, researchers Bányai and team evaluated the problematic use by the Bergen Social Media Addiction Scale (BSMAS) followed by assessing the extent to which depressive symptoms were present by the Center of Epidemiological Studies Depression-Scale (CES-D). 3 classes were the outcome of the analyses: 70.7% of the total sample were "no-risk", 15.5% were 'low risk' of problematic use while 4.5% of the social media users were the 'at-risk' group. This third group displayed elevated depressive symptoms compared to the other two classes.

Fit indices for the Latent Profile Analysis (LPA) of the social media use

Model	Log-likelihood	Replicated log-likelihood	Nr. of free parameters	AIC	BIC	SSABIC	Entropy	LMR-LRT test	*p*
2 classes	-43837	Yes	19	87711	87837	87778	0.96	12838	<0.0001
3 classes	**-42241**	**Yes**	**26**	**84534**	**84708**	**84626**	**0.94**	**3140**	**<0.05**
4 classes	-41097	Yes	33	82260	82481	82376	0.95	2251	0.69

Note: AIC = Akaike Information Criterion, BIC = Bayesian Information Criterion, SSABIC = sample size adjusted BIC, LMR-LRT = Lo–Mendell–Rubin Likelihood Ratio Test. Bold data indicate that the three-class solution was selected as a result of the LPA analysis.

doi:10.1371/journal.pone.0169839.t001

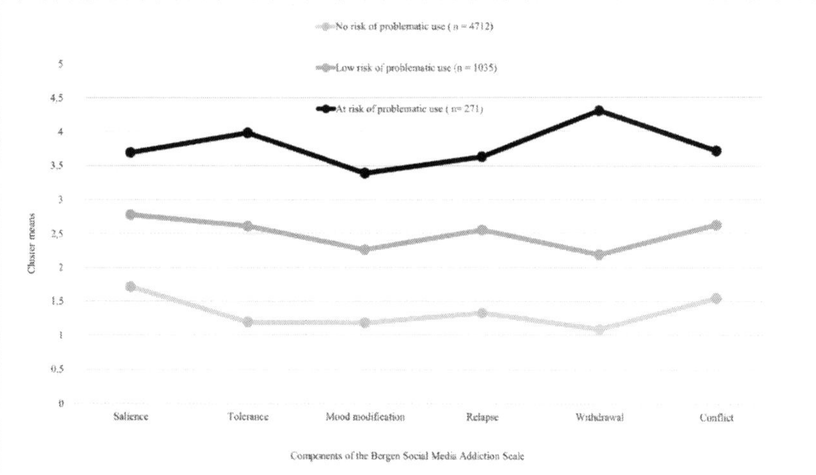

	No risk class (n = 4712)	Low risk class (n = 1035)	At-risk class (n = 271)	Overall test	
				Wald χ^2	*p* value
Gender (male %)	50.36_a	44.51_b	41.2_b	13.58	0.001
Age (years); Mean (SE)	$16.60\ (0.02)_a$	$16.61\ (0.03)_a$	$16.69\ (0.06)_a$	1.81	0.405
Weekly internet use (min 0.5, max 42 hours, mean 23.49, SD 12.73); Mean (SE)	$22.12\ (0.19)_a$	$27.11\ (0.43)_b$	$31.49\ (0.81)_c$	168.06	0.001
Weekly social media use (min 0.5. max 42 hours, mean 23.13, SD 15.56); Mean (SE)	$21.38\ (0.23)_a$	$27.68\ (0.48)_b$	$33.73\ (0.83)_c$	264.26	0.001
Self-esteem (min 1, max 4, mean 2.73, SD 0.61); Mean (SE)	$2.79\ (0.01)_a$	$2.54\ (0.02)_b$	$2.44\ (0.04)_c$	155.10	0.001
Level of depressive symptoms (min 1, max 4, mean 1.93, SD 0.60); Mean (SE)	$1.85\ (0.01)_a$	$2.163\ (0.02)_b$	$2.36\ (0.05)_c$	210.12	0.001

Note: Different subscript letters (a, b, c) in the same row reflect significant (p< 0.05) difference between the means while same subscript letters in one row reflect non-significant difference between the means according to pair wised Wald χ^2 test of mean equality for latent class predictors in mixture modeling (www.statmodel.com/download/meantest2.pdf).

doi:10.1371/journal.pone.0169839.t002

Figure 2.2: Above summarizes the results of the higher depressive symptoms in adolescents (16 years old) who were problematic social media users (Bányai et al, 2017: This is an open-access article distributed under the terms of the Creative Commons Attribution License, which permits unrestricted use, distribution, and reproduction in any medium, provided the original author and source are credited).

With this background data, let us look at the connection of depression with social media. This becomes significant given that the Royal Society for Public Health, & Young Health Movement (2017) estimated that over 25 years, the occurrence of depression has increased by 70% in adolescents!

According to McCrae et al (2018), their systematic review of Medline, PsycInfo, and Embase databases yielded eleven eligible studies of how social media usage was associated with depression in adolescents. The net conclusion was "a small but statistically significant correlation between social media use and depressive symptoms".

Table 2.1: below **TOP**: summarizes the studies analyzed of the association of depression and social media usage in adolescents; **BOTTOM**: significant findings(McCraeet al, 2018: Open Access: This article is distributed under the terms of the Creative Commons Attribution 4.0 International License (http://creativecommons.org/licenses/by/4.0/), which permits unrestricted use, distribution, and reproduction in any medium, provided you give appropriate credit to the original author(s) and the source, provide a link to the Creative Commons license, and indicate if changes were made): *p<0.01; **p<0.001

Authors (year)	Country	Design	Age	Overall sample	Girls (%)	Mental/ behavioural outcome(s)	Instrument for depressive symptoms
Ybarra et al. (2005)	USA	Cross-sectional	10–17	1501	47	Depressive symptoms	Nine symptoms from DSM
van den Eijnden et al. (2008)	Netherlands	Longitudinal cohort	12–15	663	52	Depression, loneliness	Depressive Mood List
Hwang et al. (2009)	Taiwan	Cross-sectional	12–17	6341	51	Mood	Daily Life and Depressive Mood Survey
Selfhout et al. (2009)	Netherlands	Longitudinal cohort	14–17	307	51	Depressive symptoms, social anxiety	Children's Depression Inventory

Authors (year)	Country	Design	Age	Overall sample	Girls (%)	Mental/ behavioural outcome(s)	Instrument for depressive symptoms
Gamez-Guadix (2014)	Spain	Longitudinal cohort	13–17	957	61	Depressive symptoms	Brief Symptom Inventory (depression subscale)
Neira and Barber (2014)	Australia	Cross-sectional	13–17	1819	55	Social self-concept, self-esteem, depressed mood	Depressed Mood Scale
Frison and Eggermont (2015)	Belgium	Cross-sectional	High school	910	52	Stress, depressed mood	Centre for Epidemiological Studies Depression Scale for Children (five items)
Nesi and Prinstein (2015)	USA	Cross-sectional	12–16	619	57	Depressive symptoms	Short Mood and Feelings Questionnaire
Tiggemann and Slater (2015)	Australia	Cross-sectional	10–13	204	100	Self-objectification, body shame, dieting, depressive symptoms	Children's Depression Inventory (short form)
Morin-Major et al. (2016)	Canada	Longitudinal cohort	12–17	88	53	Cortisol levels, stress, self-esteem, depressive symptoms	Children's Depression Inventory
Dumitrache et al. (2012)	Romania	Cross-sectional	16–17	123	68	Self-image, depression	Beck's Depression Inventory

Authors (year)	Included sample	Correlates	Correlation result
Ybarra et al. (2005)	1061	Internet used most frequently for chatroom versus other purposes/depressive symptoms	0.14**
Van der Eijnden et al. (2008)	663	Instant messaging (time 1)/depression (time 2)	0.17**
Hwang et al. (2009)	6341	Online communication/depressed mood	0.13*
Selfhout et al. (2009)	307	Instant messaging (time 1)/depression (time 2)	−0.02
Neira and Barber (2014)	1819	Social media use/depressive symptoms	−0.09*
Gámez-Gaudix (2014)	699	Preference for online social interaction (time 1)/depression (time 2)	0.13*
Frison and Eggermont (2015)	910	Social support seeking on Facebook/ depressed mood	0.13**
Nesi and Prinstein (2015)	619	Technology-based social comparison and feedback-seeking (time 1)/depressive symptoms (time 2)	0.34**
Tiggemann and Slater (2015)	204	Social media use/depressive symptoms	0.19*
Morin-Major et al. (2016)	88	Facebook use/depressive symptoms	−0.097
Dumitrache et al. (2012)	76	Items of identity-related information on Facebook/depression	0.355*

Another systematic review published in 2020 assessed depression arising due to social media usage in young adults (13 to 18 years of age). 13 papers were selected from 6598 articles from databases Medline, Embase, PsychINFO, Cumulative Index to Nursing and Allied Health (CINAHL) and Social Sciences Citation Index (SSCI). Analyses revealed complex relations involving multiple factors (opening the avenue of future research options) where a correlation was revealed between depression and 4 aspects of using social media: time spent; activity; investment; and addiction (Keles et al, 2020).

Table 2.2: Below summarizes the various studies assessed exploring how social media use is a double-edged sword presented here in the context of mental health (BetulKeles, Niall McCrae & Annmarie Grealish, 2020, Published by Informa UK Limited, trading as Taylor & Francis Group: This is an Open Access article distributed under the terms of the Creative Commons Attribution License (http://creativecommons.org/licenses/by/ 4.0/), which permits unrestricted use, distribution, and reproduction in any medium, provided the original work is properly cited)

Study	Aim	Design	Sample size	Sample characteristics	Outcome measure(s)	Findings/ results
O'Dea and Campbell (2011)	To explore the effect of online interaction on psychological distress	Cross-sectional	400	Aged 13–16 54.8% female	The Kessler (K-6) Scale (Kessler et al., 2003)	Negative correlation between the time spent on SM and psychological distress
Dumitrache et al. (2012)	To emphasize the relations between depression and the identity outlined on FB	Cross-sectional	76	Aged 16–17 68.4% female	Beck Depression Inventory (Beck, Ward, Mendelson, Mock, &Erbaugh, 1961)	Significant correlations between depression and the number of identity related pieces of information on FB
Neira and Barber (2014)	To investigate the relationship between social media use and depressed mood	Cross-sectional	1819	Aged 13–17 55% female	Depressed Mood Scale (Cronbach's $\alpha = 0.76$)	No meaningful relationship between SM use frequency & depressed mood. A positive association with investment in SM and depressed mood

Tsitsika et al. (2014)	To investigate the associations between heavier SNS use and internalizing behaviours.	Cross-sectional	10,930	Aged 14–17 52.3% female	Youth Self Report (YSR) problem checklist (Achenbach &Resorta, 2001)	A positive association between heavier SM use (more than 2 h/day) and internalizing problems (anxiety and depression).
Hanprathet et al. (2015)	To investigate the relationship between Facebook addiction and mental health	Cross-sectional	832	Mean age = 16.7 years SD = 1.0 Grade 10th–12th 63.5% female	The Thai General Health Questionnaire (GHQ-28; Goldberg & Williams, 1988)	A positive association between Facebook addiction and depression
Sampasa-Kanyinga and Lewis (2015)	To explore the relationship between SNSs use and psychological distress	Cross-sectional	753	Mean age = 14.1 years Grades 7th–12th 48.5% female	The Kessler (K-10) Scale (Kessler et al., 2002, Kessler et al., 2003)	The use of SNSs more than 2 hours per day was related to increased level of psychological distress.
Banjanin et al. (2015)	To investigate potential relationship between internet addiction and depression	Cross-sectional	336	Mean age = 18 years 66% female	Center for Epidemiologic Studies of Depression Scale for Children (Faulstich, Carey, Ruggiero, Enyart, & Gresham, 1986)	No relationship between time spent on SM and depression or between SM activities and depression.

Frison and Eggermont (2016)	To provide a deeper understanding of the relationships between different types of Facebook use, perceived online social support, and boys' and girls' depressed mood	Cross-sectional	910	Mean age = 15.44 years SD = 1.71 51.9% female	Center for Epidemiological Studies Depression Scale for Children (Faulstich et al., 1986)	Positive correlation between passive FB use and depressed mood as well as between active FB use and depressed mood Perceived online social support mediated this relationship; and gender influenced this association.
Vernon et al. (2017)	To examine change in problematic social networking investment and disrupted sleep, in relation to change in depressed mood and externalizing behaviour	Cohort	874	Mean age = 14.4 years Grade 9th and 11th 59% female	Depressed mood scale (adapted from the longitudinal Michigan Study of Adolescent Life Transitions) (Barber, Eccles, & Stone, 2001)	Increased investment in SM predicted higher depressed mood in adolescents, which was explained by the impact of higher levels of sleep disruptions.
Barry, Sidoti, Briggs, Reiter, and Lindsey (2017)	To determine the relations between adolescent social media use and adolescents psychosocial adjustment	Cross-sectional	226 (113 adolescent-parent dyads),	Aged 14–17 45.1% female N = 7 (6.2% did not report gender)	DSM checklist V (APA, 2013)	SM activity (#of accounts, frequency of checking) was moderately, positively correlated with anxiety and depression as reported by parents.

Li et al. (2017)	To examine the mediating effects of insomnia on the associations between online social networking addiction and depression	Cross-sectional	1015	Grade 7th–9th 41.2% female	Chinese version of the Center for Epidemiological Studies Depression scale (Chen, Yang, & Li, 2009)	A significant association between SM addiction and depression and insomnia partially mediated this relationship.
Yan et al. (2017)	To determine the time spent on SNSs, and its association with anxiety	Cross-sectional	2625	13–18 years 46.9% female	Middle School Student Mental Health Scale to measure anxiety (developed by Wang, 2006)	A positive association between time spent on SM and the level of anxiety. More than 2 hours/day and &anxiety level
Wang et al. (2018)	To examine whether rumination mediated the relation between SNS addiction and depression, and whether the mediating effect was moderated by self-esteem	Cross-sectional	365	Age 14–18 year (M_{age} = 15.96; SD = 0.69) 52% female	CES-D scale for depression (Radloff, 1977)	SM addiction and depression were positively associated. Rumination mediated this relationship and self-esteem moderated this mediation

Karim et al (2020) searched the words (1) "social media", (2) "mental health", (3) "social media" AND "mental health", (4) "social networking" AND "mental health", and (5) "social networking" OR "social media" AND "mental health" on Google scholar to focus on finally 16 selected research papers. It was reported that depression was an outcome of using social media unnecessarily. Another cause is the usage of social media for sexual fun. An important observation in many studies was depression due to the constant necessity to maintain one's self-image on social media.

Table 2.3: Below displays few studies assessed for how social media "pressure" influences depressive symptoms(Karim et al, 2020: This is an open-access article distributed under the terms of the Creative Commons Attribution License, which permits unrestricted use, distribution, and reproduction in any medium, provided the original author and source are credited): IGU, internet gaming disorder; PSMU, problematic social media use.

Author	Title of Study	Method	Findings
Berryman et al. [10]	Social Media Use and Mental Health among Young Adults	Cross-sectional	Social media use was not predictive of impaired mental health functioning.
Coyne et al. [11]	Does Time Spent using Social Media Impact Mental Health?: An Eight Year Longitudinal Study	8-year longitudinal study	Increased time spent on social media was not associated with increased mental health issues across development when examined at the individual level.
Escobar-Viera et al. [12]	For Better or for Worse? A Systematic Review of the Evidence on Social Media Use and Depression Among Lesbian, Gay, and Bisexual Minorities	Systematic Literature Review	Social media provides a space to disclose minority experiences and share ways to cope and get support; constant surveillance of one's social media profile can become a stressor, potentially leading to depression.
O'Reilly et al. [15]	Potential of Social Media in Promoting Mental Health in Adolescents	qualitative study	Adolescents frequently utilize social media and the internet to seek information about mental health.
O'Reilly [16]	Social Media and Adolescent Mental Health: The Good, the Bad and the Ugly	focus groups	Much of the negative rhetoric of social media was repeated by mental health practitioners, although there was some acknowledgement of potential benefit.

Feder et al. [17]	Is There an Association Between Social Media Use and Mental Health? The Timing of Confounding Measurement Matters	longitudinal	Frequent social media use report greater symptoms of psychopathology.
Rasmussen et al. [19]	The Serially Mediated Relationship between Emerging Adults' Social Media Use and Mental Well-Being	Exploratory study	Social media use may be a risk factor for mental health struggles among emerging adults and that social media use may be an activity which emerging adults resort to when dealing with difficult emotions.
Keles et al. [19]	A Systematic Review: The Influence of Social Media on Depression, Anxiety and Psychological Distress in Adolescents	systematic review	Four domains of social media: time spent, activity, investment, and addiction. All domains correlated with depression, anxiety and psychological distress.
Nereim et al. [21]	Social Media and Adolescent Mental Health: Who You Are and What You do Matter	Exploratory	Passive social media use (reading posts) is more strongly associated with depression than active use (making posts).
Mehmet et al. [22]	Using Digital and Social Media for Health Promotion: A Social Marketing Approach for Addressing Co-morbid Physical and Mental Health	Intervention	Social marketing digital media strategy as a health promotion methodology. The paper has provided a framework for implementing and evaluating the effectiveness of digital social media campaigns that can help consumers, carers, clinicians, and service planners address the challenges of rural health service delivery and the tyranny of distance,

Odgers and Jensen [23]	Adolescent Mental Health in the Digital Age: Facts, Fears, and Future Directions	Review	The review highlights that most research to date has been correlational, has focused on adults versus adolescents, and has generated a mix of often conflicting small positive, negative, and null associations.
Twenge and Martin [24]	Gender Differences in Associations between Digital Media Use and Psychological Well-Being: Evidence from Three Large Datasets	Cross-sectional	Females were found to be addicted to social media as compared with males.
Fardouly et al. [25]	The Use of Social Media by Australian Preadolescents and its Links with Mental Health	Cross-sectional	Users of YouTube, Instagram, and Snapchat reported more body image concerns and eating pathology than non-users, but did not differ on depressive symptoms or social anxiety
Wartberg et al. [26]	Internet Gaming Disorder and Problematic Social Media Use in a Representative Sample of German Adolescents: Prevalence Estimates, Comorbid Depressive Symptoms, and Related Psychosocial Aspects	Cross-sectional	Bivariate logistic regression analyses showed that more depressive symptoms, lower interpersonal trust, and family functioning were statistically significantly associated with both IGD and PSMU.
Neira and Barber [28]	Social Networking Site Use: Linked to Adolescents' Social Self-Concept, Self-Esteem, and Depressed Mood	Cross-sectional	Higher investment in social media (e.g. active social media use) predicted adolescents' depressive symptoms. No relationship was found between the frequency of social media use and depressed mood.

In today's world where competition and stress along with compromised family relations make an individual seek solace on social media. While social media offers positive benefits such as to reach out to someone in distress or in times of danger to "online" interactions there is the pressure of "maintaining the image" to being taken for a ride by "online" romances or friendships.

The occasional "selfie" post is fine as long as we know where to strike the balance and enjoy the benefits offered by social media to quell depression rather than get more depressed by using it!

2.2 COVID-19 AND DEPRESSION

Viral outbreaks have been part of history with the Spanish Flu of the early 20th century, Severe Acute Respiratory Syndrome (SARS) in 2003, H1N1 influenza in 2009, etc. While the world was busy planning for the New Year in December 2019, pneumonia was reported in Wuhan, China. Later named as the new Coronavirus (COVID-19), the condition was declared as a pandemic. The symptoms range from fever to sore throat to chills to cough to muscle pain to vomiting, and diarrhea. In severe forms, acute respiratory syndrome and death are the outcomes. Apart from these physical manifestations, the psychological outcomes of the outbreak ranged from depression to fear to helplessness. Such health crises are associated with anxiety, fear, and depression among the citizens as well as medical workers. Further studies of those who were affected displayed manifestations such as mood swings, depression, anger, stress, distress, insomnia, and attention deficit hyperactivity disorder(Salari et al, 2020).

Depression and mental illness have been conspicuously elevated after mass traumatic events. Depressive symptoms were reported by 9.6% of Manhattan residents post-September 11, 2001. The 2019 Hong Kong civil protests were accompanied by reports of increased depression and anxiety: 6.5% prior to the unrest to 11.2% during the unrest. Financial recessions that adversely affected the income or social standing have also been associated with increased mental illness.

The SARS and Ebola outbreaks were associated with higher rates of mental illness. On the same lines, COVID-19 too has had its share of depression. 50.4% of health care workers in China who were exposed to patients with COVID-19 reported depressive symptoms. From the start to mid-April 2020, 20 million people filed for unemployment in the US!

(Ettman et al, 2020)

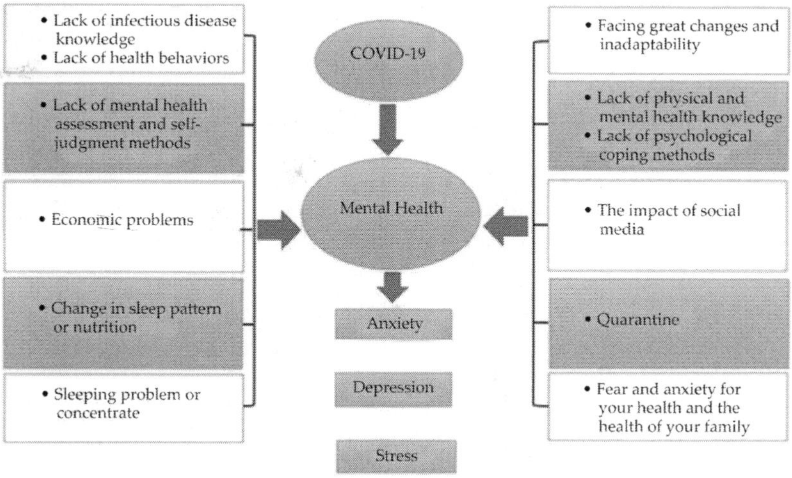

Figure 2.3: Above displays the various aspects of mental health challenges by the COVID-19 pandemic (Salariet al, 2020: This is an open-access article distributed under the terms of the CC-BY license, which permits unrestricted use, distribution, and reproduction in any medium. You are not required to obtain permission to reuse this article content, provided that you credit the author and journal).

Researchers Lai and team looked at healthcare workers of the COVID-19 pandemic and mental health employing demographic data and mental health measurements from 1257 health care workers in 34 hospitals from January 29, 2020, to February 3, 2020, in China. Analyses revealed the participants to be 60.8% nurses, 39.2% physicians, 60.5% workers in Wuhan hospitals, and 41.5% frontline health care workers. 50.4% of the participants reported depression while distress was recorded in 71.5% of the participants. More severe scores of mental health symptoms were reported by nurses, women, frontline health care workers, and those working in Wuhan. The risk of symptoms for depression was revealed to be (OR, 1.52; 95% CI, 1.11-2.09; P=.01) along with anxiety (OR, 1.57; 95% CI, 1.22-2.02; P<.001), insomnia (OR, 2.97; 95% CI, 1.92-4.60; P<.001), and distress (OR, 1.60; 95% CI, 1.25-2.04; P<.001) in frontline health care professionals.

Table 1. Demographic and Occupational Characteristics of Responders

	No. (%)					
		Occupation		Location		
Characteristic	Total	Physician	Nurse	Wuhan	Hubei province outside Wuhan	Outside Hubei province
Overall	1257 (100)	493 (39.2)	764 (60.8)	760 (60.5)	261 (20.8)	236 (18.8)
Sex						
Men	293 (23.3)	223 (45.2)	70 (9.2)	146 (19.2)	52 (19.9)	95 (40.3)
Women	964 (76.7)	270 (54.8)	694 (90.8)	614 (80.8)	209 (80.1)	141 (59.7)
Age, y						
18-25	198 (15.8)	10 (2.0)	188 (24.6)	162 (21.3)	32 (12.3)	4 (1.7)
26-30	407 (32.4)	126 (25.6)	281 (36.8)	258 (33.9)	111 (42.5)	38 (16.1)
31-40	406 (32.3)	200 (40.6)	206 (27.0)	224 (29.5)	71 (27.2)	111 (47.0)
>40	246 (19.5)	157 (31.8)	89 (11.6)	116 (15.3)	47 (18.0)	83 (35.2)
Marriage status						
Unmarried	418 (33.3)	87 (17.6)	331 (43.3)	314 (41.3)	66 (25.3)	38 (16.1)
Married[a]	839 (66.7)	406 (82.4)	433 (56.7)	446 (58.7)	195 (74.7)	198 (83.9)
Education level						
≤Undergraduate	953 (75.8)	217 (44.0)	736 (96.3)	611 (80.4)	238 (91.2)	104 (44.1)
≥Postgraduate	304 (24.2)	276 (56.0)	28 (3.7)	149 (19.6)	23 (8.8)	132 (55.9)
Technical title						
Junior	699 (55.6)	153 (31.0)	546 (71.5)	481 (63.3)	169 (64.8)	49 (20.8)
Intermediate	378 (30.1)	187 (37.9)	191 (25.0)	221 (29.1)	61 (23.4)	96 (40.7)
Senior	180 (14.3)	153 (31.1)	27 (3.5)	58 (7.6)	31 (11.8)	91 (38.5)
Place of residence						
Urban	1220 (97.1)	474 (96.1)	746 (97.6)	751 (98.8)	247 (94.6)	222 (94.1)
Rural	37 (2.9)	19 (3.9)	18 (2.4)	9 (1.2)	14 (5.4)	14 (5.9)
Working position						
Frontline	522 (41.5)	176 (35.7)	346 (45.3)	390 (51.3)	72 (27.6)	60 (25.4)
Second-line	735 (58.5)	317 (64.3)	418 (54.7)	370 (48.7)	189 (72.4)	176 (74.6)
Type of hospital						
Tertiary	933 (74.2)	369 (74.8)	564 (73.8)	538 (70.8)	218 (83.5)	177 (75.0)
Secondary	324 (25.8)	124 (25.2)	200 (26.2)	222 (29.2)	43 (16.5)	59 (25.0)

[a] Married category included widowed and divorced participants.

Figure 2.4: Above summarizes the details of how depression was reported by front line health workers in a China-based study (Lai et al, 2020: This is an open-access article distributed under the terms of the CC-BY license, which permits unrestricted use, distribution, and reproduction in any medium. You are not required to obtain permission to reuse this article content, provided that you credit the author and journal):

In another nationally representative survey study, researchers Ettman and team explored depression symptoms among US adults before and during the COVID-19 pandemic. The study involved employing the National Health and Nutrition Examination Survey, conducted from 2017 to 2018 before COVID-19 and the COVID-19 and Life Stressors Impact on Mental Health and Well-being study, conducted from March 31, 2020, to April 13, 2020, was the source of data for during the epidemic.

The participant details (5065 participants) before the epidemic were (1704 participants [37.8%] aged 18-39 years; 2588 [51.4%] women; 1790 [62.9%] non-Hispanic White) while for during the epidemic: 1441 participants (619 participants [43.0%] aged 18-39 years; 723 [50.2%] men; 933 [64.7%] non-Hispanic White).

Assessment on the basis of the Patient Health Questionnaire–9 revealed that 27.8% of the participants had depression during the epidemic vs. 8.5% prior to the COVID-19 epidemic. In the words of the researchers, “In this population-representative survey study of US adults, we found that prevalence of depression symptoms was more than 3-fold higher during COVID-19 compared with the most recent population-based estimates of mental health in the US. This increase in depression symptom prevalence is higher than that recorded after previous mass traumatic events, likely reflecting the far more pervasive influence of COVID-19 and its social and economic consequences than other, previously studied mass traumatic events.” (Ettman et al, 2020)

Looking at the factors associated with depression, the symptoms of depression were more in women than men; those who were married had a lower prevalence of depression vs. those who were widowed/divorced/single/with a partner. The final factor was resources with lower depression in those with more resources: depression was observed in 46.9% of those who had the lowest income vs. 16.9% in the highest income group.

Figure 2.5 below displays the spike in depression symptoms in US adults in a representative survey study during the COVID-19 epidemic vs. before the epidemic. Factors influencing depression were income and marital status(Ettman et al, 2020: This is an open-access article distributed under the terms of the CC-BY license, which permits unrestricted use, distribution, and reproduction in any medium. You are not required to obtain permission to reuse this article content, provided that you credit the author and journal):

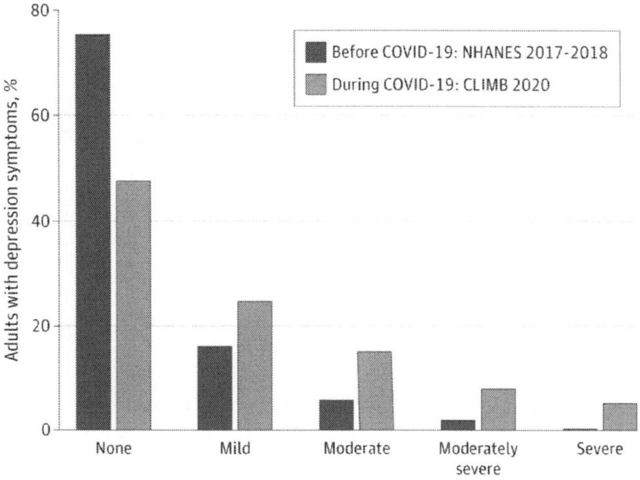

Table 2. Prevalence of Depression Symptoms in US Adults Before and During the COVID-19 Pandemic

	% (95% CI)[b]		Difference	
Depression symptoms[a]	**Before COVID-19[c]**	**During COVID-19[d]**	**Absolute, %**	**Relative**
None	75.3 (73.3-77.1)	47.5 (44.2-50.9)	-27.7	0.6
Mild	16.2 (15.1-17.4)	24.6 (21.8-27.7)	8.4	1.5
Moderate	5.7 (4.8-6.9)	14.8 (12.6-17.4)	9.1	2.6
Moderately severe	2.1 (1.6-2.8)	7.9 (6.3-9.8)	5.7	3.7
Severe	0.7 (0.5-0.9)	5.1 (3.8-6.9)	4.4	7.5

Abbreviation: COVID-19, coronavirus disease 2019.

[a] Depression symptoms assessed using the Patient Health Questionnaire-9 and categorized as none (score, 0-4), mild (score, 5-9), moderate (score, 10-14), moderately severe (score, 15-19), and severe (score, ≥20).

[b] Percentages weighted to the noninstitutionalized US adult population.

[c] Before COVID-19 estimates from the National Health and Nutrition Examination Survey from 2017 to 2018.

[d] During COVID-19 estimates from the COVID-19 and Life Stressors Impact on Mental Health and Well-being study collected from March 31, 2020, to April 13, 2020.

Table 3. Odds of Depression Symptoms by Resources and Exposure to COVID-19 Stressors

Resource	OR (95% CI)[a]	*P* value
Education		
<High school	0.86 (0.41-1.77)	.68
High school graduate or GED	1.58 (0.96-2.60)	.08
Some college	1.45 (0.97-2.16)	.07
≥College	1 [Reference]	NA
Marital status		
Married	1 [Reference]	NA
Widowed, divorced, or separated	2.08 (1.29-3.36)	.003
Never married	1.85 (1.17-2.94)	.008
Living with partner	1.27 (0.74-2.18)	.39
Household income, $		
≤19 999	2.37 (1.26-4.43)	.007
20 000-44 999	1.32 (0.78-2.24)	.31
45 000-74 999	1.04 (0.63-1.72)	.88
≥75 000	1 [Reference]	NA
Household savings, $		
≤4999	1.52 (1.02-2.26)	.04
≥5000	1 [Reference]	NA
COVID-19 stressors		
Low	1 [Reference]	NA
Medium	1.77 (1.16-2.71)	.008
High	3.05 (1.95-4.77)	<.001

Abbreviations: COVID-19, coronavirus disease 2019; GED, general education diploma; NA, not applicable; OR, odds ratio.

[a] Complete case analysis used for multiple logistic regression resulting included 1386 participants for this model. Model controls for demographic characteristics (ie, sex, age, race/ethnicity, and household size). Depression symptoms defined as PHQ-9 score cutoff of 10 or greater. COVID-19 stressor score calculated from stressor summation ranging from 0-13; categories represent low (score, 0-2), medium (score, 3-4), and high (score, 5-13) exposure to stressors due to COVID-19. Data on household income were missing for 34; on household savings, 45 participants; and COVID-19 stressor score, 3 participants.

Figure 2.5: Above displays the spike in depression symptoms in US adults in a representative survey study during the COVID-19 epidemic vs. before the epidemic. Factors influencing depression were income and marital status(Ettman

According to a systematic review and meta-analysis by Salariand team(2020), of 17 cross-sectional studies of depression, anxiety, and COVID-19, the prevalence of depression was found to be 33.7% in the general population. It was also discussed that following COVID-19 news can increase the levels of anxiety given the extent of rumors. It becomes ethical for the media to publish news on the recovery or new treatment approaches to lower mental stress. Though the risk of mortality due to COVID-19 is more in the aged, depression and stress were more in the 21–40-year-old population. This is attributed to social media access and fears of career/the economic future.

Table 2.4: Below displays the prevalence of depression and other mental health issues based on a systematic review and meta-analysis (Salari et al, 2020: This is an open-access article distributed under the terms of the CC-BY license, which permits unrestricted use, distribution, and reproduction in any medium. You are not required to obtain permission to reuse this article content, provided that you credit the author and journal).

Psycho-logical disorders	continents	Num-ber of articles	Sample Size	I^2 (%)	Egger's test	Prevalence (95% CI)
anxiety	Asia	13	54,596	99.2	0.136	32.9 (95% CI:28.2–37.9)
	Europe	3	8341	98.8	0.272	23.8 (95% CI:16.2–33.5)
depres-sion	Asia	10	35,688	99.5	0.224	35.3 (95% CI:27.3–44.1)
	Europe	3	8341	99.2	0.104	32.4 (95% CI:21.6–45.5)
stress	Asia	3	2758	96.3	0.229	27.9 (95% CI:19.7–37.8)
	Europe	2	6316	98.5	–	31.9 (95% CI:23.1–42.2)

In a nutshell, this data is suggestive of the importance of taking mental health issues and addressing them during major traumatic events such as viral outbreaks.

2.3 PREGNANCY AND DEPRESSION

While the incidence of depression in women is 20%, the vulnerability to feeling blue rises during pregnancy. Considered the most prevalent psychiatric disorder during pregnancy, the prevalence range is 9% to 28% and 25% to 50% for predominantly middle class and low-income populations, respectively. Several factors are involved to manifest as depression during this tender period of a woman's life ranging from gender-based violence to insufficient nutrition to poor care to economic challenges to social issues toearlier psychiatric issues to previous pregnancy-associated issues such as abortions or stillbirths or complications (Ajinkya et al, 2013).

In order to address the issue, researchers FekaduDadi and team published a current systematic review and meta-analysis in 2020 exploring the extent of antenatal depression along with risk factors, and its effect on delivery (protocol number CRD42017082624 on PROSPERO). This employed querying terms "Pregnan*, antenat*, depression, clinical depression, depressed mood, major depressive disorder, depressive symptom, adverse birth outcomes, stillbirth, preterm birth, and low birth weight" in the following databases: CINHAL, MEDLINE, EMCare, PubMed, Psych INFO and, Psychiatry online, and Scopus. Following quality assessment, data synthesis, and analyses 73 articles were narrowed down upon of which 64 dealt with antenatal depression and 9 articles were the effects of antenatal depression on birth outcomes.

The occurrence of depression displayed an increase over the 3 trimesters: 17.1% (95%CI: 7.7%-26.5%) in the first trimester, 27.1% (95%CI: 19.7%-34.6%) in the second trimester, and 28.9% (95%CI: 23.7%-34.1%) in the third trimester. Compared against high-income nations and community-based studies, the incidence of depression was more in low-income countries (PP = 34.1%, 95%CI: 22.7%-45.6%) and in health institution-based studies (PP = 27.6%, 95%CI: 23%-32.3%). Risk factors for depression during pregnancy were bad obstetric history (Pooled Odds Ratio (POR) = 2.01; 95%CI: 1.67, 2.42) followed by financial stress (POR = 2.03; 95%CI: 1.63, 2.53), poor social support (POR = 1.77; 95%CI: 1.49, 2.10), ahistory of common mental disorders (POR = 3.27; 95%CI: 2.47, 4.33), history of violence (POR = 2.99; 95%CI: 2.20, 4.07), relationship issues (POR = 2.18; 95%CI: 1.64, 2.90), and preference for male gender (POR = 1.41; 95%CI: 2.97; 6.26).

The risk of adverse birth outcomes was 59% more due to antenatal depression. This was attributed to stress hormone production (for example, cortisol) that could target adequate food/oxygen supply to the fetus or smoking/drinking habits in depressed mothers to higher risks of infection in mothers due to weakened immune responses, on account of depression.

Variables of sub-analysis	Number of studies (%)	Sample size	Pooled prevalence; 95%CI
Year of publication			
2007–2008	2(3.13)	375	9.46 (6.5–12.5)
2009–2010	8(12.50)	4,448	26.4(19.5–33.4)
2011–2012	16(25.00)	9,533	28.6(22.3–34.8)
2013–2014	12(18.75)	8,116	23.8(18.3–29.3)
2015–2016	20(31.25)	11,194	26.8(17.8–35.8)
2017	6(9.38)	10,369	18.2(14.6–21.8)
Income of the country			
Low income	15(23.44)	8346	34.1(22.7–45.6)
Middle income	49(76.56)	35,689	22.7(20.1–25.2)
Study setting			
Health institution	46(71.88)	26536	27.6(22.9–32.3)
Community based	18(28.13)	17499	19.8(16.1–23.5)
Time of screening			
First trimester	4(6.25)	5170	17.1 (7.7–26.5)
Second trimester	8(12.50)	9912	27.1(19.7–34.6)
Third trimester	20(31.25)	9532	28.9(23.7–34.1)
All trimester	32(50.00)	19421	23.9(18.2–29.5)
Median sample size			
< = 600	42(65.62)	13290	25.7(22.0–29.5)
>600	22(34.38)	30745	24.8(18.9–30.6)
Tool used for screening depression			
EPDS	34(53.13)	23,612	27.2(23.5–30.8)
CIS-R	2(3.13)	578	19.3(13.8–24.9)
PHQ-9	6(9.38)	6157	25.4(17.3–33.4)
SRQ-20	2(3.13)	2669	15.4(4.4–26.4)
Hopkins symptom checklist	2(3.13)	1740	58.9(20.9–96.8)
BDI	6(9.38)	1580	22.2(15.7–28.6)
HADS	7(10.94)	5829	18.6(11.8–25.4)
DSM-IV	2(3.13)	289	27.4(10.5–65.3)
CESD-10	3(4.69)	1581	13.6(11.9–15.3)

BDI: Beck Depression Inventory **EPDS**: Edinburgh Postnatal Depression scale **HADS**: Hospital Anxiety and Depression Scale **CIS-R**: Clinical Interview Schedule Revised **SRQ**: Self Reporting Questionnaire **CESD-10**: Center for Epidemiological Studies Depression Scale **DSM-V**: Diagnostic and Stastical Manual of Mental Disorder

https://doi.org/10.1371/journal.pone.0227323.t002

Variable of sub-analysis	Number of studies	Sample size	POR, 95%CI	I^2, p-value
Poor obstetric history (*history of adverse birth outcome, unwanted pregnancy, obstetric complications*)	16	13450	2.01(1.67,2.42)	81.7%, p = 0.137
Economic difficulties	14	11207	2.03(1.63,2.53)	74.3%, p = 0.001
Poor social support	13	7372	1.77(1.49,2.10)	85.7%, p = 0.001
History of CMD(*depression, anxiety, stressful life events*)	13	11799	3.27(2.47,4.33)	89.9%, p = 0.001
History of all forms of violence	11	7428	2.99(2.20, 4.07)	71.7%, p = 0.001
Unsatisfied marital condition (*Unmarried, divorced, separated, shorter marital duration, polygamous*)	9	7533	2.18(1.64,2.90)	73.0%, p = 0.001
Male gender preference (the family preferred male than girl)	4	1135	2.97(1.41,6.26)	88.2%, p = 0.001

https://doi.org/10.1371/journal.pone.0227323.t003

Author, Year	Country, income	Study setting	Study design	Sample size	Follow up start time	Tool used for screening	LBW (<2500gm) Estimate (RR/OR)	PB(<37weeks), Estimate (OR)
Rahman A et al, 2007[104]	Pakistan, Low	Community	Prospective cohort	290	3rd	ICD-10	1.9;1.3–2.9)	
Nasreen HE et al 2010[105]	Bangladesh, Middle	Community	Prospective cohort	720	2nd and 3rd	EPDS> = 10	2.24, 1.37–3.68	
Niemi M et al, 2013[106]	Vietnam, Middle	Community	Prospective cohort	334	3rd	EPDS > = 3	2.40;1.09–5.25	2.07, 1.2–3.56
Sanchez SE et al, 2013[107]	Peru, Middle	HI	Case control	959	3rd	PHQ-9> = 10		3.67, 2.09–6.46
Chang HY et al, 2014[26]	Korea, Low	HI	Prospective cohort	691	3rd	CESD-10> = 10	1.66; 0.55–5.02	
Husain N et al, 2014[27]	Pakistan, Low	Community	Prospective cohort	763	3rd	EPDS > = 12	0.88; 0.73–1.06	
Rao D et al, 2015 [108]	India, Middle	HI	Prospective cohort	150	2nd & 3rd	PHQ-9> = 5		3.3, 0.99–11.17
Bindt C et al 2013[69]	Ghana, Middle	HI	Longitudinal, birth cohort	719	3rd	PHQ-9 > = 10	β = 52.2; 18.2–122.6	2.1, 0.8–5.6
Wado WD et al 2014[109]	Ethiopia, Low	Community	Longitudinal, birth cohort	537	2nd & 3rd	EPDS> = 13	1.77; 1.03–3.04	

LBW: Low Birth weight HI: Health Institutions **ICD-10**: International classification of Disease 10th **EPDS**: Edinburgh Postnatal Depression Scale **SRQ**: Self Reporting Questionnaire **CESD-10**: Center for Epidemiological Studies Depression Scale **DSM-V**: Diagnostic and Stastical Manual of Mental Disorder **PHQ**: Patient Health Questionnaire

https://doi.org/10.1371/journal.pone.0227323.t004

Figure 2.6: Above displays the prevalence of depression in pregnant women in low/middle-income countries followed by the risk factors to depression, and the association of this depression with birth outcomes (FekaduDadi et al, 2020: This is an open-access article distributed under the terms of the Creative Commons Attribution License, which permits unrestricted use, distribution, and reproduction in any medium, provided the original author and source are credited).

Another systematic review (protocol number CRD42018116267 on PROSPERO) assessed studies on antenatal depression (January 1st, 2007 and August 31st, 2018) employing the Assessment of Multiple Systematic Reviews (AMSTAR) checklist scores. The researchers who described the publication as "the first systematic review of reviews examining antenatal depression and adverse birth outcomes published to date" concluded that "Globally, antenatal depression prevalence was high and could be considered a common mental disorder during pregnancy". The prevalence of antenatal depression was found to be 15 to 65%, assessed from (306 studies with 877,246 participants) on the prevalence of antenatal depression and (39 studies with 75,451 participants) for preterm and low birth weight due to antenatal depression. The major risk factors were identified as earlier or current abuse or violence, reduced or absent social or partner support, unplanned pregnancy, history of obstetric complications, financial challenges, and lower educational status.

Table 2.5: Below displays the risk factors of depression in pregnancy as assessed in the systematic review (Dadi et al, 2020: Open Access: This article is distributed under the terms of the Creative Commons Attribution 4.0 International License (http://creativecommons.org/licenses/by/4.0/), which permits unrestricted use, distribution, and reproduction in any medium, provided you give appropriate credit to the original author(s) and the source, provide a link to the Creative Commons license, and indicate if changes were made. The Creative Commons Public Domain Dedication waiver (http://creativecommons.org/publicdomain/zero/1.0/) applies to the data made available in this article unless otherwise stated):

Risk factors	Number of reviews in which the risk factor was reported	Number of primary studies in which the factor was reported	Total participants
History of abuse (childhood or current sexual, physical or psychological) or domestic violence or intimate partner violence	6	73	293,621
Lack of partner or of social support and poor marital relationship	4	47	226,078
Personal or family history of any mental disorder or stress	3	34	177,014
Un-planned or unwanted pregnancy specially during premarital condition or nullparity	3	36	70,296
History of poor obstetric condition like current or past pregnancy complications such as hyperemesis gravidurum, adverse birth outcomes (low birth weight, preterm, still birth or infant lose after delivery), had cesarean section delivery	4	33	56, 916

Maternal low economic status or unemployment condition or financial difficulties	3	32	20,239
Maternal poor behavioral condition or practices like smoking, alcohol use, illicit drug use)	1	29	18,444
Maternal low educational status	2	22	14,638
	10 reviews	306 primary studies	877,246 participants

Further pooling and analyses of data revealed that the risk of low-weight birth was 1.49 (95%CI: 1.32, 1.68; I^2 =0.0%) and 1.40 for preterm birth infants (95%CI: 1.16, 1.69; I^2 =35.2%) in depressed mothers when compared to non-depressed mothers (Dadi et al, 2020).

"A mother's joy begins when new life is stirring inside… when a tiny heartbeat is heard for the very first time, and a playful kick reminds her that she is never alone."

– Unknown

Given that becoming a mother ushers in mixed feelings of both joy and stress, the occurrence of depression during pregnancy is being highlighted by research to be "high". This warrants the incorporation of appropriate public health policies to address the risk factors and thereby allow a woman to enjoy the journey of becoming a mother.

2.4 HOW HAPPY ARE WE?

"Mental illness is one of the main causes of unhappiness. This is not a tautology (…) for people can be unhappy for many reasons from poverty to unemployment to family breakdown to physical illness" (World Happiness Report, 2013).

Given that the quality of life is different when one is happy, interestingly the emotional well-being scores were found to be moderately high in people with mental disorders according to the Adult Psychiatric Morbidity Survey for England 2007. Researchers Spinhoven and team (2015) presented data that extraversion (obtaining gratification from outside oneself) and neuroticism (a tendency to be negative) are the most vital of the Big Five personality traits to determine being happy. These researchers further assessed cross-sectional as well as longitudinal relationships of happiness in relation to depression and anxiety (scored by self and a clinician).

2142 adults aged 18–65:healthy controls and persons with past or current depressive and/or anxiety disorder constituted the sample. Measures of happiness, depression, and anxiety were assessed at the time pointsT0 (year 0), T2 (year 2), and T4 (year 4) in order to predict happiness and emotional well-being at T6 (year 6). The measures were inclusive of the Self-Rating of Happiness scale (SRH) for happiness, the Mood and Anxiety Symptom Questionnaire—Shortened Dutch 30-item version (MASQ-D30) for emotional health, the Composite Interview Diagnostic Instrument (CIDI, version 2.1) for various psychiatric disorders, Inventory of Depressive Symptomatology self-report version (IDS-SR) for depressive symptom scores, and the NEO Five-Factor Inventory (NEO-FFI) for neuroticism and extraversion.

24.3% of the sample reported to be depressed: Interesting outcomes emerged from the analyses that are summarized below:

Education was associated with the scores for differences in happiness (SRH) and emotional well-being (EWB) while age was associated with the former.

15.5% of depressed individuals were 'quite unhappy', 'very unhappy' or 'completely unhappy' while 47 % of those who had both social anxiety disorder with major depressive disorder reported to feeling so.

The authors state "we found that the overall affective appraisal of being happy and the level of emotional well-being gradually decreased from controls to persons with a past anxiety/depressive disorder, to persons with a current single anxiety or single depressive disorder to persons with a current comorbid anxiety/depressive disorder".

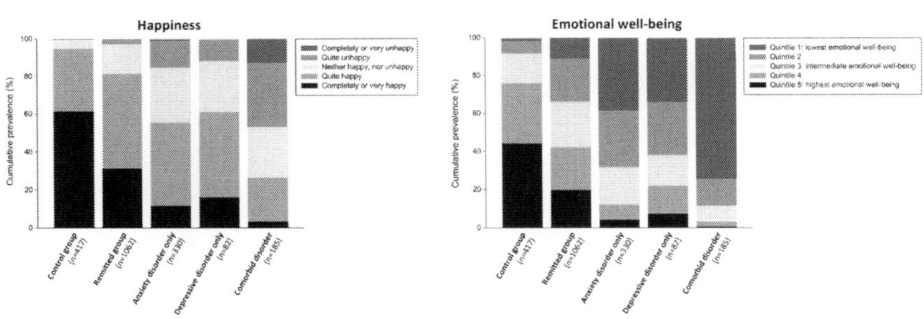

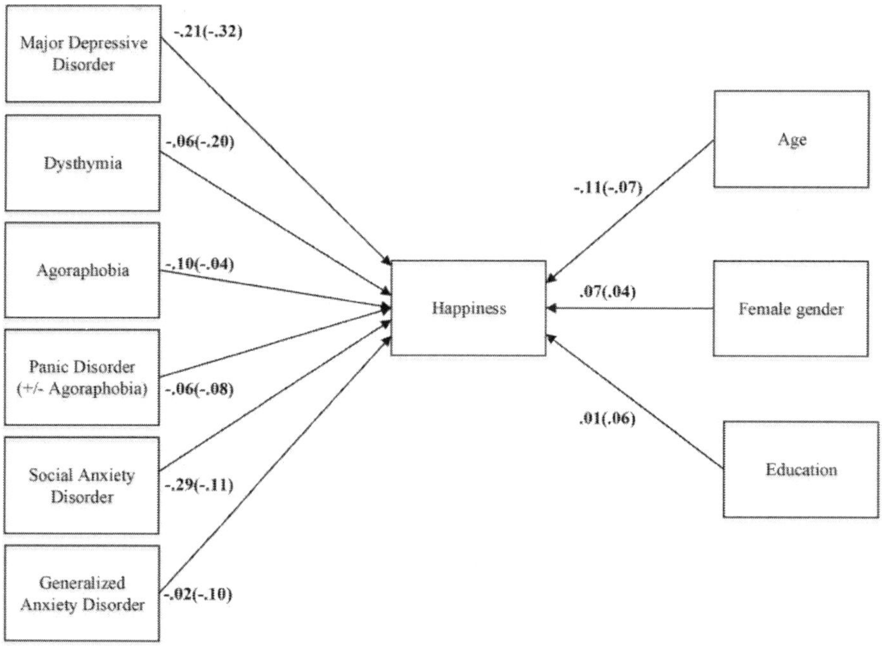

Figure 2.7: Above displays **TOP**: happiness and emotional well-being across healthy and depressed individuals; **BOTTOM**: factors associated with "happiness" in the Netherlands Study of Depression and Anxiety (NESDA)-based study(Spinhoven et al, 2015: This is an open-access article distributed under the terms of the Creative Commons Attribution License, which permits unrestricted use, distribution, and reproduction in any medium, provided the original author and source are properly credited):

In a nutshell, the study outlined that happiness and emotional well-being were reported to be lower in people with depressive symptoms or emotional disorders. This becomes significant for addressing such concerns by health caregivers to boost positive feelings as opposed to quelling depressive symptoms in people who say they are "not born to be happy".

(Spinhoven et al, 2015)

2.5 LOOKING AT DEPRESSION IN ADOLESCENTS

According to the WHO Fact Sheet (28 September 2020), 50% of mental health issues originate by 14 years of age but, detection and treatment are often neglected. The sheet further states that among adolescents, depression accounts for the major cause of disability and illness. According to Grover et al (2019), the prevalence of depression in adolescents was 3%–68% in school-based studies and 0.1%–6.94% in community studies.

According to Bernaras, Jaureguizar and Garaigordobil (2019), child and adolescent depression can be evaluated by many approaches. The Children's Depression Inventory-CDI based on Beck's cognitive theory of depression is ubiquitously utilized. Another system was designed called the Children's Depression Scale-CDS that screens for negative affectivity. The Depression Self-Rating Scale for Children-DSRS looks at any compromised trait in a child or adolescent that impedes his/her ability to perform well. Another one is the Short Mood and Feelings Questionnaire (SMFQ) that also looks at restlessness and feeling tired.

Table 2.6: Below shows tests to diagnose the child and adolescent depression- **TOP**: self-administered; **BOTTOM**: Hetero-administered (Bernaras, Jaureguizar and Garaigordobil, 2019: This is an open-access article distributed under the terms of the **Creative Commons Attribution License (CC BY)**. The use, distribution or reproduction in other forums is permitted, provided the original author(s) and the copyright owner(s) are credited and that the original publication in this journal is cited, in accordance with accepted academic practice. No use, distribution or reproduction is permitted which does not comply with these terms)

Name	References	Num. items	Dimensions (variables measured)	Target age group	Psychometric characteristics	Spanish adaptation
Children's Depression Scale (CDS)	Lang and Tisher, 1978	66	Depressive total: Affective Response, Social Problems, Self-esteem, Pre-occupation with own Sickness or Death, Guilt and Various Depressives. Positive Total: Mood-Joy and Various Positives.	8–16	K-R20 = 0.91; Alpha = 0.92–0.94; Guttman split-half coefficient = 0.90; Test-retest reliability = 0.74 (Bath and Middleton, 1985; Tisher et al., 1992)	Seisdedos, 2003
Children's Depression Inventory (CDI)	Kovacs, 1985	27	Scales: emotional problems, functional problems. Subscales: negative mood/physical symptoms, negative self-esteem, interpersonal problems, ineffectiveness	7–17	Alpha = 0.75–0.90; κ = 0.76; K-R20 = 0.80–0.94; Test-retest reliability = 0.62–0.82; Sensitivity = 0.81–0.95; Specificity = 0.64–0.96; PPV = 0.21–0.90; NPPV = 0.63–1.00 (view review by Stockings et al., 2015)	Del Barrio and Carrasco, 2004
Center for Epidemiological Studies Depression Scale for Children (CES-DC)	Weissman et al., 1980	20	Total depression	12–18	Alpha = 0.86–0.89; Sensitivity = 0.71–0.82; Specificity = 0.62–0.90; PPV = 015; NPV = 0.96 (view review by Stockings et al., 2015)	Soler et al., 1997
Depression Self-Rating Scale for Children (DSRS)	Birleson, 1981	18	Total depression	8–14	Alpha = 0.86 Test-retest reliability = 0.80; Sensitivity = 0.67; Specificity = 0.77; PPV = 0.15; NPV = 0.97 (Birleson et al., 1987)	No
Reynolds Adolescent Depression Scale (RADS)	Reynolds, 1987	30	Total depression, dysphoric mood, Anhedonia/negative affect, Negative self-evaluation, somatic complaints	13–17	Alpha = 0.92 (view review by Stockings et al., 2015); Sensitivity = 0.86; Specificity = 0.49; PPV = 0.69; NPV = 0.72 (Krefetz et al., 2002)	Del Barrio et al., 1996
Reynolds Child Depression Scale (RCDS)	Reynolds, 1989	30	Total depression	7–13	Alpha = 0.85–0.91 Test-retest reliability = 0.82–0.85 (Reynolds, 1989); Sensitivity = 0.79; Specificity = 0.87 (Figueras et al., 2008)	Figueras et al., 2008
Mood and Feelings Questionnaire (MFQ), and Short Mood and Feelings Questionnaire (SMFQ)	Angold et al., 1995	32 (MFQ) 13 (SMFQ)	Total depression	8–18	MFQ: Alpha = 0.90–0.93; AUC = 0.86 (95% CI: 0.81, 0.91), Sensitivity = 0.84; Specificity = 0.70. SMFQ: Alpha = 0.87–0.89; AUC = 0.86 (95% CI: 0.80, 0.91); Sensitivity = 0.84; Specificity = 0.68 (Thabrew et al., 2017)	No
Beck Depression Inventory (BDI-II)	(Beck et al., 1996)	21	Total depression	13 and over	Alpha = 0.92–0.94; Sensitivity = 0.74–0.88; Specificity = 0.70–0.92; PPV = 0.76–0.85; NPV = 0.67–0.95 (view review by Stockings et al., 2015)	Sanz and Vázquez, 1998
Revised Child Anxiety and Depression Scale—RCADS	Chorpita et al., 2000	Two versions: with 47 and 30 items	Anxiety and separation disorders. Social phobia, Generalized anxiety, Panic, Obsessive compulsive disorder and Major depressive disorder. Total Anxiety and Depression Score	8-18	Alpha = 0.78; sensitivity = 0.74; specificity = 0.77 for the Major Depression scale (MDD). RCADS MDD scale correlated positively and significantly with the CDI	RCADS: (Sandín et al., 2009); RCADS-30: (Sandín et al., 2010)
Kutcher Adolescent Depression Scale (KADS)	Leblanc et al., 2002	16 (long), 11 (short), and 6 (brief)	Total depression	6–18	KADS-16 Alpha = 0.82; AUC = 0.85 KADS-11: Alpha = 0.84; AUC = 0.94, (95%CI: 0.91, 0.97); Sensitivity = 0.89; Specificity = 0.90 KADS-6: Alpha = 0.80 Sensitivity = 0.92 Specificity = 0.71; PPV = 0.10 for total sample and.26 for clinical sample; NPV = 1.0 for the total sample and.99 for the clinical sample (Leblanc et al., 2002; Brooks et al., 2003; Zhou et al., 2016)	There is a Spanish version, but no data are available regarding its validation.
Cuestionario Educativo Clínico de ansiedad y depresión (*Clinical Educational Anxiety and Depression Questionnaire*) (CECAD)	Lozano et al., 2007	50	Depression, anxiety, worthlessness, irritability, problems with thinking and psycho-physiological symptoms	7-adulthood	Alpha = > 0.83 Omega coefficient = 0.77–0.87. Correlations with CDI between 0.26 and 0.76.	–

KR-20, Kuder-Richardson coefficient (formula 20); κ, Cohen's kappa reliability co-efficient; PPV, Positive predictive value; NPV, Negative predictive value; AUC, Area under the Receiver Operating Characteristic Curve (AUC).

TABLE 4 | Hetero-administered psychometric tests for assessing child and adolescent depression.

Name	References	Num. items	Dimensions (variables measured)	Administered by	Target age group	Psychometric characteristics	Spanish adaptation
Children's Depression Rating Scale-Revised (CDRS-R)	Poznanski et al., 1979	17	Total depression	Clinical personnel (interviews with child and parents)	6–12	Alpha = .85	No
Escala para la evaluación de la depresión para maestros (*Teachers' scale for evaluating depression*) (ESDM)	Domènech-Llaberia and Polaino-Lorente, 1990	16	Performance, social interaction, inhibited depression, and anxious depression	Teachers	8–12	Alpha = 0.88	–
Diagnostic Interview for Children and Adolescents–Revised (DICA-R)	Reich et al., 1991	1–2 h	Disruptive behavior disorders, mood disorders, anxiety disorders, eating disorders, and elimination disorders	Clinical personnel (interviews with child and parents)	8–18	High intra-rater reliability and moderate agreement between parents-children/adolescents (see Ezpeleta et al., 1995).	Ezpeleta et al., 1995
Semistructured Clinical Interview for Children and Adolescents (SCICA)	McConaughy and Achenbach, 1994	224	Anxiety, depression, motor/language problems, attention problems, self-control problems, aggression, somatic complaints	Clinical personnel	6–18	Mean test-retest reliability: 0.78	No
The Kiddie Schedule for Affective Disorders and Schizophrenia-Present and Lifetime (K-SADS-PL)	Kaufman et al., 1996	82+5 modules	Major depression, dysthymia, mania, hypomania, cyclothymia, bipolar disorder, schizoaffective disorder, schizophrenia, schizophreniform disorder, brief reactive psychosis, panic disorder, agoraphobia, separation anxiety disorder, avoidant disorder of childhood and adolescence, simple phobia, social phobia, generalized anxiety, obsessive compulsive disorder, attention deficit hyperactivity disorder, conduct disorder, oppositional defiant disorder, enuresis, encopresis, anorexia nervosa, bulimia, transient tic disorder, Tourette's disorder, chronic motor or vocal tic disorder, alcohol abuse, substance abuse, post-traumatic stress disorder, and adjustment disorders	Clinical personnel	6–18	High convergent validity and limited divergent validity (Lauth et al., 2010).	No
Diagnostic Interview Schedule for Children (DISC)	Shaffer et al., 2000	1–2 h	Anxiety disorders, mood disorders, disruptive disorders, substance abuse, others (anorexia, bulimia, enuresis/encopresis, selective mutism, schizophrenia, etc.)	Clinical personnel (interviews with child and parents)	6–17	No validity data available. Acceptable test-retest reliability (Shaffer et al., 2000)	Bravo et al. 2001
Screening de problemas emocionales y de conducta infantil (*Screening for emotional and behavioral problems in children*) (SPECI)	Garaigordobil and Maganto, 2012	10 min	Withdrawal, somatization, anxiety, infantile-dependent, problems with thinking, attention-hyperactivity, disruptive behavior, academic performance, depression, and violent behavior	Teachers	5–12	Alpha = 0.82	–

In order to address this trend of depression in adolescents, screening for prevention programsat the schoollevel. An example is a cognitive-behavioral therapy in depressed adolescents as part of the Penn Prevention program.The utility of family-based treatment also facilitates the alleviation of depression and personal growth issues in depressed people.

Table 2.7: Below shows programs of depression at the school-level (Bernaras, Jaureguizar and Garaigordobil, 2019: This is an open-access article distributed under the terms of the **Creative Commons Attribution License (CC BY)**. The use, distribution or reproduction in other forums is permitted, provided the original author(s) and the copyright owner(s) are credited and that the original publication in this journal is cited, in accordance with accepted academic practice. No use, distribution or reproduction is permitted which does not comply with these terms)

Name	References	Target age group	Aims	Type	Characteristics	Results
Penn Resiliency Program (PRP)	Gillham et al., 1990	8–15	To raise awareness of the relationship between cognition, emotion and behavior, help youngsters develop social and decision-making skills and foster optimism	U	- Cognitive-behavioral perspective - 12 sessions, each lasting 90–120 min - Contents: cognitive restructuring, maladaptive coping strategies, attributional styles, assertiveness, negotiation, relaxation, decision-making, social skills	Significant reduction in depression levels assessed using the CDI (Gillham, 1994; Gillham et al., 1995; Quayle et al., 2001; Chaplin et al., 2006; Cardemil et al., 2007) as well as in anxiety levels and behavioral problems, and fostering of wellbeing and optimism. Other universal (Pattison and Lynd-Stevenson, 2001; Gillham et al., 2007) and targeted studies (Gillham and Reivich, 1999; Roberts et al., 2003a, 2004) failed to find any evidence that the program had any impact on the variables analyzed, in either the short or long term. The targeted study by Gillham et al. (2006) found that PRP did not significantly prevent depressive disorders but significantly prevented depression, anxiety, and adjustment disorders (when combined) among high-symptom participants
Coping with Stress Course (CWSC)	Clarke et al., 1995	13–17	To challenge irrational thoughts, cope with negative moods, overcome passivity, and reach agreements with parents and peers; social skills training	T	- Target population: Adolescents with some known increased risk of depression: past episode of depression; persistent sub-diagnostic dysphoria and/or other depressive symptoms; depressed parents; pregnant, single teen mother; other known risk factors for depression - Based on Beck's (Beck et al., 1979) and Ellis' (Ellis and Harper, 1961) cognitive therapy - 15 sessions lasting 45 min and 8 sessions lasting 90 min - Contents: depression and its relationship with stress, training in cognitive restructuring skills and the modification of irrational thoughts	Significant reduction in depression levels and the risk of reappearance at posttest and during follow-up (8, 12, and 18 months) (Clarke et al., 1995, 2001; Garber et al., 2009) (depressive symptoms assessed by the Center for Epidemiologic Studies-Depression Scale-CES-D). Participants who were currently on antidepressant medication were excluded from these studies
Aussie optimism program	Rooney et al., 2000	6–11	To intervene in risk and protection factors for depression and anxiety (cognitive, emotional, and social characteristics).	U	- Cognitive-behavioral perspective - 10 sessions, each lasting 60 min - Contents: identifying negative beliefs about oneself and one's present and future circumstances, identifying and regulating emotions, engaging in pleasurable activities, working with hierarchies of situations which generate fear and learning to relax (Rooney et al., 2013)	- Rooney et al. (2006) applied the program to 72 children aged between 8 and 9 and found both a significant reduction in depression levels (assessed using the CDI) and more positive causal attributions at posttest and during short-term follow up, although these results were not found to persist in the long term. The effect sizes observed were low ($\eta^2 = 0.09$ for the symptoms of depression and $\eta^2 = 0.16$ for attributional style). - The program was also found to reduce depressive symptoms (assessed using the CDI) in a study with 47 female 7th grade students, with a 6-month follow up (Quayle et al., 2001)

Resourceful Adolescent Program-Adolescents (RAP-A)	Shochet et al., 2001	12–15	To identify and challenge irrational thoughts, provide training in social skills and problem solving and help prevent conflicts with parents and peers	U	- Cognitive-behavioral perspective - 11 sessions, each lasting 40–50 min - Contents: cognitive restructuring, problem-solving, social skills, and communication training	Significant results in preventing depression in random groups at posttest (measured using the BDI-II and RADS) but not during follow up, at least according to the BDI-II (the effect persisted according to the RADS) (Merry et al., 2004), although when non-random groups were used the results were also significant during follow up (Shochet et al., 2001)
FRIENDS	Barrett and Turner, 2001	7–16	To reduce the incidence of anxiety and depression, emotional distress and social problems, teaching children how to cope with anxiety, both now and in the future	U	- Cognitive-behavioral perspective - 10 sessions, each lasting 120 min + 2 booster sessions - Contents: one's own and other people's emotions, relaxation, trying to do your best, planning steps, making time to have fun together, friendship and family skills, being happy, etc.	Reduced anxiety levels, although the results regarding reduced depression levels are more limited (Barrett and Turner, 2001; Lowry-Webster et al., 2001; Lock and Barrett, 2003). In Barrett and Turner's study with 489 children aged between 10 and 12, the effect size for depression (measured using the CDI) at posttest was 0.09. Curiously enough, self-reported depression decreased slightly in the psychologist-led condition, while slight increases were noted in the teacher-led condition, although these increases were statistically, but not clinically, significant (Barrett and Turner, 2001)
Problem Solving for Life (PSFL)	Spence et al., 2003	13–15	Cognitive restructuring, problem solving	U	- Cognitive-behavioral perspective - 8 sessions, each lasting 45–50 min - Contents: challenging maladaptive thoughts, coping with problems	No significant results were found as regards preventing depressive symptoms (Spence et al., 2005; Sheffield et al., 2006), although an improvement was observed in all the variables studied in all study groups (those that did and those that did not participate in the program) (Sheffield et al., 2006)
Interpersonal Psychotherapy-Adolescent Skills Training (IPT-AST)	Young and Mufson, 2003	11–16	Training in social skills, coping with life transitions, and overcoming interpersonal deficits	T	- Target population: adolescents with elevated symptoms of depression - Cognitive-behavioral perspective - 10 sessions, each lasting 90 min - Contents: social and communication skills	Immediate reduction in depressive symptoms, although the benefits did not persist longer than 6 months (Horowitz et al., 2007; Young et al., 2010) (depressive symptoms assessed using the Center for Epidemiologic Studies-Depression Scale-CSD-D and the Children's Depression Rating Scale (CDRS)
Adolescents Coping with Emotions (ACE)	Sheffield et al., 2006	14–15	To prevent or reduce depression levels, improving coping skills, and fostering resilience	T	- Target population: adolescents with elevated symptoms of depression (those scoring in the top 20% on the combined scores, sum of standardized scores, the Children's Depression Inventory (CDI) and the Center for Epidemiologic Studies—Depression Scale (CES-D). - Cognitive-behavioral perspective - Applied by teachers - 8 sessions, each lasting 60 min - Contents: cognitive restructuring, training in social skills and assertiveness, negotiation and problem-solving skills, recognition of one's own achievements	Significant reduction in depressive symptoms and negative thoughts in girls after 6 months (Kowalenko et al., 2005), but not during the 12-month follow up (Sheffield et al., 2006). Depressive symptoms were assessed using the CDI and CES-D
FORTIUS	Méndez et al., 2013	8–13	To psychologically strengthen participants at a cognitive, emotional and behavioral level	U	- 12 sessions lasting 50–60 min +2 booster sessions one month later +1 final session 3 months later - Contents: understanding and controlling negative emotions, social skills, and the ability to organize everyday activities, detection and restructuring of negative thoughts, problem solving, decision making, and positive self-instruction	No significant differences were found in depression (measured using the CDI) at posttest, although a reduction was observed during follow up (12 months) in girls (Orenes, 2015)

The treatment of depression in adolescents should involve multiplefactors ranging from the individual patient along with the type of relationship with their families and friends. This opens the avenue of addressing depression should involve not only the school environment but also their home and family.

2.6 THE SWEET DISEASE AND DEPRESSION:

According to the International Diabetes Federation, "diabetes is one of the largest global health emergencies of the 21st century". When compared to individuals with normal glucose metabolism, the occurrence of depression is moderately increased in prediabetic patients and in diabetic patients who are yet to be diagnosed while there is an evident increase in diagnosed individuals. Research points out a bidirectional association: with the prevalence of depression higher by a factor of two and three in type 2 diabetes and with type 1 diabetes, respectively. The risk of type 2 diabetes is more by 60% in depressed individuals. This warrants accurate analyses given that the prognosis and compliance to treatment are challenged in diabetic individuals due to depression (Bădescu et al 2016).

In order to assess the risk factors associated with depression in diabetic individuals, researchers Tusa and team (2020)reported a hospital-based cross-sectional study employingBayesian logistic regression models on a dataset at Adama Hospital and Medical College (AHMC), Ethiopia. The measures applied on 359 diabetic patients (average age: 51 (±14.51) years) were the Kessler 6 scales for depression, sociodemographic, and clinical factors along with adherence to medication and drinking habits. Analyses revealed that 9.22% of these patients were depressed: with a higher prevalence in urban settings and those who had a history of being admitted to the hospital, had complications, and did not adhere to prescribed medication regimens. The risk of depression was 29.29 and 24.46 times more in those who did not adhere to medication and those who took oral antidiabetic medication and insulin against those who took only oral antidiabetic medication. The factors that could strongly predict depression in diabetic patients were higher sugar levels, complications, admission to the hospital, and failure to adhere to prescribed medication.

Table 2.8: Below displays **TOP**: features of the patients; **BOTTOM**: the factors estimated to be predictors of depression in diabetic patients: DM: diabetes mellitus; HPD: highest posterior density; MC: Monte Carlo; AOR: adjusted odds ratio; SD: standard deviation (BirukShalmenoTusa et al, 2020: This is an open access article distributed under the Creative Commons Attribution License, which permits unrestricted use, distribution, and reproduction in any medium, provided the original work is properly cited).

Variable	**Frequency (n=359)**	**Percentage (%)**
Hazardous drinking		
Yes	22	6.13
No	337	93.87
Type of DM		
Type 1	47	13.09
Type 2	312	86.91
Types of medication		
Oral	221	61.56
Insulin	110	30.64
Both	28	7.80
Medication adherence		
Low	101	28.13
Medium	86	23.96
High	172	47.91
DM complication		
Yes	83	23.12
No	276	76.88
Chronic illness other than DM		
Yes	147	40.95
No	212	59.05
Hospital admission		
Yes	104	28.97
No	235	71.03

A thorough comprehension of the association of both the "Ds": depression and diabetes can allow for efficient screening and handling of both the conditions and make life "sweeter".

2.7 DEPRESSION: MULTIPLE PATHWAYS AND SYNERGY

Research has pointed out several factors: genetic and environmental (adversity or life events or education or body mass index) to account for gene-environment interactions (G × E) to influence major depressive disorder (MDD). With this background, researchersHullam and team (2020) utilized probabilistic graphical models (Bayesian networks) to assess the interdependenciesofhealthfactors:social,environmental,lifestyle,metabolic, and mental on depression. The hierarchy of higher-order interactions was assessed employing Bayesian multilevel analysis of relevance (BMLA) on the basis of the UK Biobank resource (110,599 subjects: based on lifetime depression).

It was revealed that direct associations with lifetime depression were seen with current depressive symptoms, neuroticism, parental depression, and body fat percentage or what is called obesity. Other factors

All other factors shown on the graph are in a non-direct relationship with reported lifetime depression, most of them mediated by current depressive symptoms: these are inclusive of sleep quality (insomnia) and pain (stomach, neck, head, shoulder, knee, general). Transitive relations were detected between lifetime depression and negative life events of the past two years, having a confidant to share events, physical activity, household income, along with consumption of tobacco and alcohol.

In a nutshell, the authors concluded that lifetime depression involves direct and indirect roles of health, environmental, lifestyle, and social aspects-the results are direct for some and transitive for others. Factors associated with current depressive symptoms influence financial, social, sporting, pain, and sleep aspects, age, the Townsend deprivation index, and satisfaction. Headache, insomnia, and social factors involve the role of neuroticism. Metabolic and sport factors are influenced by body fat. Income, alcohol consumption, maternal smoking, stress, taking risks: these aspects are influenced by body fat, and parental depression.

The use of artificial intelligence (known for its emerging roles in a broad spectrum of fields: Shiv and Prerna, in press): here a deep neural network-based classifier outlined the ability of a few of these factors to function as predictors of the condition. The factors that displayed significant values were neuroticism, current depressive symptomsparental depression,and body fat.

Interestingly, these factors also showed synergy among them: lifetime depression was influenced by high neuroticism and high current depressive symptoms(OR: 9.26 and OR: 10.66 respectively)-a combination of these 2 factors with parental depression and high body fat displayed a large multivariate odds ratio.

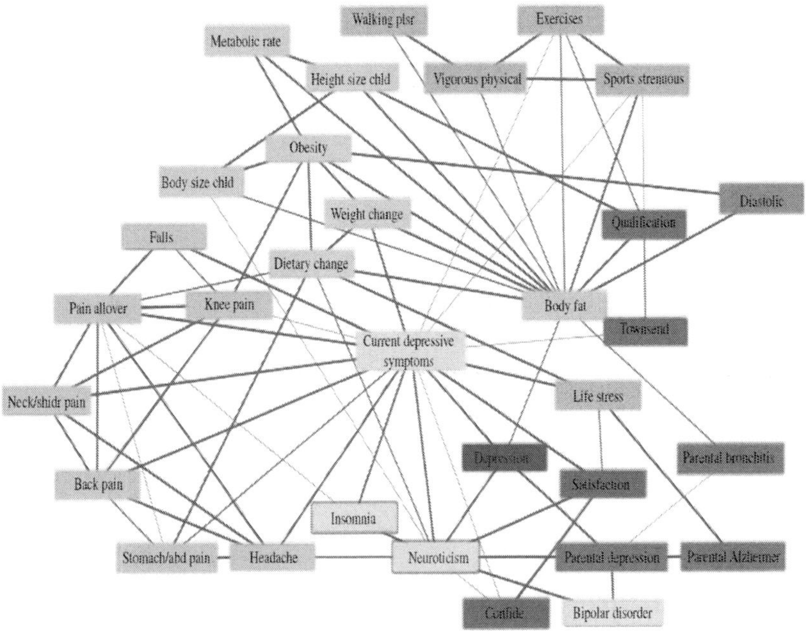

Below: higher-order interactions that influence lifetime depression are variables connected with the same marker; the probability of relevance of a variable is the height of the red column for the variable:

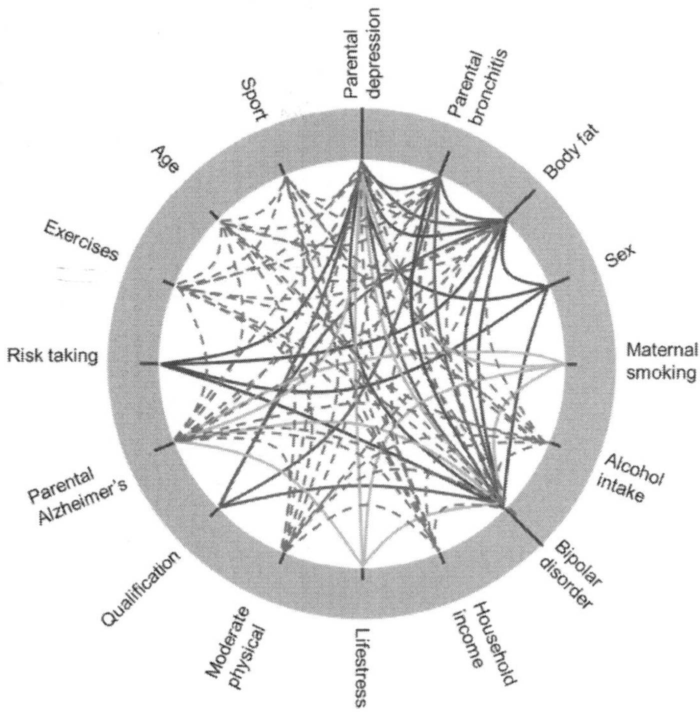

.		I.+II.		I-II.+III.		I-III.+IV.			
Neuroticism		**Parental depression**		**Body fat**		**Current depressive symptoms**			
	CR-OR		**CR-OR**		**CR-OR**		**CR-OR**	**CI$_{95\%}$**	
								Low	**High**
Low	0.17	No	0.16	Normal		Low	0.16	0.14	0.19
						Moderate	0.26	0.23	0.31
					0.19	High	0.87	0.63	1.21
				High		Low	0.19	0.16	0.21
						Moderate	0.43	0.38	0.48
					0.28	High	1.50	1.22	1.85
		Yes	0.82	Normal		Low	0.47	0.32	0.67
						Moderate	0.61	0.43	0.86
					0.61	High	2.50	1.40	4.47
				High		Low	0.66	0.49	0.89
						Moderate	1.24	0.98	1.57
					1.02	High	2.69	1.66	4.35

Mod-erate	1.25	No	1.08	Normal		Low	0.55	0.44	0.68
						Moderate	0.68	0.59	0.78
					0.77	High	1.68	1.39	2.03
				High		Low	0.76	0.63	0.91
						Moderate	1.11	1.00	1.23
					1.35	High	2.79	2.47	3.16
		Yes	2.33	Normal		Low	1.30	0.82	2.04
						Moderate	1.72	1.32	2.23
					1.83	High	3.17	2.12	4.74
				High		Low	2.04	1.37	3.03
						Moderate	1.85	1.46	2.35
					2.72	High	5.86	4.54	7.57
High	5.60	No	3.93	Normal		Low	1.00	0.75	1.34
						Moderate	1.31	1.14	1.49
					2.30	High	3.93	3.55	4.34
				High		Low	1.72	1.38	2.15
						Moderate	2.15	1.94	2.38
					4.13	High	5.93	5.49	6.4
		Yes	6.37	Normal		Low	2.26	1.30	3.94
						Moderate	3.08	2.46	3.86
					4.88	High	7.26	6.12	8.62
				High		Low	2.77	1.71	4.50
						Moderate	3.89	3.19	4.74
					6.96	High	10.11	8.84	11.57

Figure 2.7: Below displays TOP chart: Bayesian map of depression and associated factors: (A) Mental health descriptors - yellow, (B) Social factors - dark green, (C) Childhood descriptors - lime, (D) Parental illnesses - brown, (E) Lifestyle and sports - light green, (F) Diet and metabolism - gold, (G) Blood pressure - orange, (H) Financial background and qualification - dark blue, (I) Pain - purple, (J) Life stress - light brown, (K) Falls - pink, and reported lifetime depression – red; BOTTOM chart: Interactions among variables showing the presence and direct interactions of neuroticism, current depressive symptoms, and parental depression. This is followed by body fat percentage ($p_{Dir}=0.800$), life stress ($p_{Dir}=0.200$) and household income; Bottom table: Neuroticism, Parental depression, Body fat, and Current depressive symptoms showing synergy with respect to reported lifetime depression

(Hullam et al, 2020: Open Access-This article is licensed under a Creative Commons Attribution 4.0 International License, which permits use, sharing, adaptation, distribution and reproduction in any medium or format, as long as you give appropriate credit to the original author(s) and the source, provide a link to the Creative Commons license, and indicate if

Thus, direct, transitive, and synergistic associations among health, environmental, lifestyle, and social aspects together bring in the environment of depression(Hullam et al, 2020).

2.8 REFERENCES

1. BetulKeles, Niall McCrae & Annmarie Grealish (2020) A systematic review: the influence of social media on depression, anxiety and psychological distress in adolescents, International Journal of Adolescence and Youth, 25:1, 79-93, DOI: 10.1080/02673843.2019.1590851
2. Karim, F., Oyewande, A. A., Abdalla, L. F., Chaudhry Ehsanullah, R., & Khan, S. (2020). Social Media Use and Its Connection to Mental Health: A Systematic Review. Cureus, 12(6), e8627. https://doi.org/10.7759/cureus.8627
3. Bányai F, Zsila Á, Király O, Maraz A, Elekes Z, Griffiths MD, et al. (2017) Problematic Social Media Use: Results from a Large-Scale Nationally Representative Adolescent Sample. PLoS ONE 12(1): e0169839. https://doi.org/10.1371/journal.pone.0169839
4. Royal Society for Public Health, & Young Health Movement. (2017). StatusOfMind social media and young people's mental health and wellbeing.
5. McCrae, N., Gettings, S. &Purssell, E. Social Media and Depressive Symptoms in Childhood and Adolescence: A Systematic Review. Adolescent Res Rev **2,** 315–330 (2017). https://doi.org/10.1007/s40894-017-0053-4
6. Salari, N., Hosseinian-Far, A., Jalali, R. et al. Prevalence of stress, anxiety, depression among the general population during the COVID-19 pandemic: a systematic review and meta-analysis. Global Health **16,** 57 (2020). https://doi.org/10.1186/s12992-020-00589-w
7. Ettman CK, Abdalla SM, Cohen GH, Sampson L, Vivier PM, Galea S. Prevalence of Depression Symptoms in US Adults Before and During the COVID-19 Pandemic. JAMA Netw Open. 2020;3(9):e2019686. doi:10.1001/jamanetworkopen.2020.19686.
8. Lai J, Ma S, Wang Y, et al. Factors Associated With Mental Health Outcomes Among Health Care Workers Exposed to Coronavirus Disease 2019. JAMA Netw Open. 2020;3(3):e203976. doi:10.1001/jamanetworkopen.2020.3976
9. Ajinkya, S., Jadhav, P. R., & Srivastava, N. N. (2013). Depression during pregnancy: Prevalence and obstetric risk factors among pregnant women attending a tertiary care hospital in Navi Mumbai. Industrial psychiatry journal, 22(1), 37–40. https://doi.org/10.4103/0972-6748.123615

10. FekaduDadi A, Miller ER, Mwanri L (2020) Antenatal depression and its association with adverse birth outcomes in low and middle-income countries: A systematic review and meta-analysis. PLoS ONE 15(1): e0227323. https://doi.org/10.1371/journal.pone.0227323
11. Dadi, A.F., Miller, E.R., Bisetegn, T.A. et al. Global burden of antenatal depression and its association with adverse birth outcomes: an umbrella review. BMC Public Health **20,** 173 (2020). https://doi.org/10.1186/s12889-020-8293-9
12. Spinhoven, P., Elzinga, B. M., Giltay, E., &Penninx, B. W. (2015). Anxious or Depressed and Still Happy?. PloS one, 10(10), e0139912. https://doi.org/10.1371/journal.pone.0139912
13. Layard R, Chisholm D, Patel VL, Saxena S. Mental illness and unhappiness In: Helliwell J, Layard R, Sachs J, editors. World Happiness Report 2013. New York: Columbia University; 2013:38–53.
14. https://www.who.int/news-room/fact-sheets/detail/adolescent-mental-health
15. Grover, S., Raju, V. V., Sharma, A., & Shah, R. (2019). Depression in Children and Adolescents: A Review of Indian studies. Indian journal of psychological medicine, 41(3), 216–227. https://doi.org/10.4103/IJPSYM.IJPSYM_5_19
16. Bernaras E, Jaureguizar J and Garaigordobil M (2019) Child and Adolescent Depression: A Review of Theories, Evaluation Instruments, Prevention Programs, and Treatments. Front. Psychol. 10:543. doi: 10.3389/fpsyg.2019.00543
17. International Diabetes Federation . IDF Diabetes. 7 ed. Brussels, Belgium: International Diabetes Federation; 2015. http://www.diabetesatlas.org. [Google Scholar] [Ref list]
18. Bădescu, S. V., Tătaru, C., Kobylinska, L., Georgescu, E. L., Zahiu, D. M., Zăgrean, A. M., &Zăgrean, L. (2016). The association between Diabetes mellitus and Depression. Journal of medicine and life, 9(2), 120–125.
19. BirukShalmenoTusa, Mekuriaw Alemayehu, AdisuBirhanuWeldesenbet, SewnetAdem Kebede, Getachew Asfaw Dagne, "Prevalence of Depression and Associated Factors among Diabetes Patients in East Shewa, Ethiopia: Bayesian Approach", Depression Research and Treatment, vol. 2020, Article ID 4071575, 11 pages, 2020. https://doi.org/10.1155/2020/4071575

20. Hullam, G., Antal, P., Petschner, P. et al. The UKB envirome of depression: from interactions to synergistic effects. Sci Rep **9,** 9723 (2019). https://doi.org/10.1038/s41598-019-46001-5
21. ShivSanjeeviSripathi and Prerna Pandey (2020). Artificial Intelligence. Delve Publications. In Press.

CHAPTER 3

THE ECONOMIC ASPECTS OF DEPRESSION

"Don't tell me where your priorities are. Show me where you spend your money and I'll tell you what they are."

—James W. Frick

CONTENTS

3.1 CASH AND DEPRESSION-RESEARCH

A study published in 2010 by researchers Birnbaum and team assessed the cost involved in treatment and the performance at work in individuals with major depressive disorder (MDD). The workplace performance along with the medication usage and costs were assessed over 12 months employing the Quick Inventory of Depressive Symptomatology Self-Report (QIDS-SR). Among 539 individuals who participated, severe MDD was reported by 47.7%; moderate by 38.5%, and mild in 13.8%. In the case of severe MDD, the average cost of treatment was $697 against $388 for the mild cases; (chi(2)=4.4, P=.019) while the cost incurred for antidepressants was $256 for the severe cases vs. $88 for the mild cases chi(2)=9.0, P=.001). Workers who were moderately and severely depressed reported more absenteeism from work as against non-depressed workers with associated loss in performance, as measured by salary was 15.7% for mildly depressed; 23.3% for moderately depressed, and 31.3% for severely depressed. The human capital costs of the loss in productivity were nearly $2 billion when projected to the US workforce. Overall, the authors concluded that "MDD severity is significantly associated with increased treatment usage/costs, treatment adequacy, unemployment, and disability and with reduced work performance" (Birnbaum et al, 2010).

Scientists Greenberg and team (2015) approximated the incremental economic burden of individuals with major depressive disorder (MDD) on the basis of national survey (DSM-IV criteria) and administrative claims data (ICD-9 codes). The authors reported that this burden was $210.5 billion its value of from $173.2 billion between 2005 and 2010, accounting for a 21.5% rise. Of the total costs incurred, MDD accounted for 38% with the rest attributed to comorbidities calling for analyzing, "further these comorbidities as well as the relative importance of factors contributing to that growing burden".

Researchers Egede and team (2016) assessed the expenditure associated with depression and diabetes utilizing the Medical Expenditures Panel Survey (MEPS). This data led to the formation of a pooled dataset over 8 years (2004–2011) of adult (n=147,095) responses. Analyses revealed that the incremental cost of depression was $2,654 against patients who did not have the condition. The authors considered this incurred cost as "significant" with potential savings possible through prompt screening of depression.

3.2 THE EFFECT OF INCOME ON DEPRESSION

On the basis of the US Panel Study of Income Dynamics (PSID) following groups of siblings and their parents for up to 40 years involving the prospective collecting of information on education, income, work, and marriage, Smith and Smith (2010) assessed the effect of childhood psychological conditions on adult life. Children who had a history of physical illness or psychological issues affected the capacity to earn in adulthood: the loss of household assets was $18,000 and an annual dip of the family income by $10,400. The situation is presented as the authors stated," The long-term economic damages of childhood psychological problems are large—a lifetime cost in lost family income of approximately $300,000, and total lifetime economic cost for all those affected of 2.1 trillion dollars".

Goodman and team (2011) explored the adult impacts in long-term of having experienced psychological and physical health problems during childhood on the basis of the British National Child Development Study where various aspects of physical and mental health (as observed by healthcare givers and questionnaires) during childhood and adolescence along with their current adult life income and health. It was observed that by the time these individuals turned 50 years of age, the participant family incomes dipped by 28%. It was presented by the researchers that," Large effects are found due to childhood psychological problems on the ability of affected children to work and earn as adults and on intergenerational and within-generation social mobility".

Scientists McMillan and team (2011) employed the Collaborative Psychiatric Epidemiology Surveys: an amalgamation of 3 nationally representative surveys of American adults conducted between 2001 and 2003 to explore the role of the household income on mental stress. After adjusting for sex, marital status, race, age, and employment status, the extent of mental distress (as measured using the Kessler Psychological Distress Scale) was the most when the income was the lowest. Among the ethnicities studied: non-Hispanic white, Hispanic, Asian American, black, and other- the first set showed a negative association between contemplating suicide and income. This according to the researchers opens up the avenue of circumventing depression in low-income, high-risk populations.

Based on a symptom-screening questionnaire, the Patient Health Questionnaire (PHQ-9), a study from the National Health and Nutrition Examination Survey examined the NHANES 2007–2016 data to unveil interesting aspects of depression. Overall, the report showed that 8.1%

of Americans, over 20 years of age were considered depressed between 2013-2016 over the two-week study period. Interestingly, the likelihood of depression was 5.5% in men vs. 10.4% in women across all the age groups analyzed. In terms of ethnicity, 9.2% of non-Hispanic black individuals analyzed were depressed vs. 3.1% of non-Hispanic Asian adults.

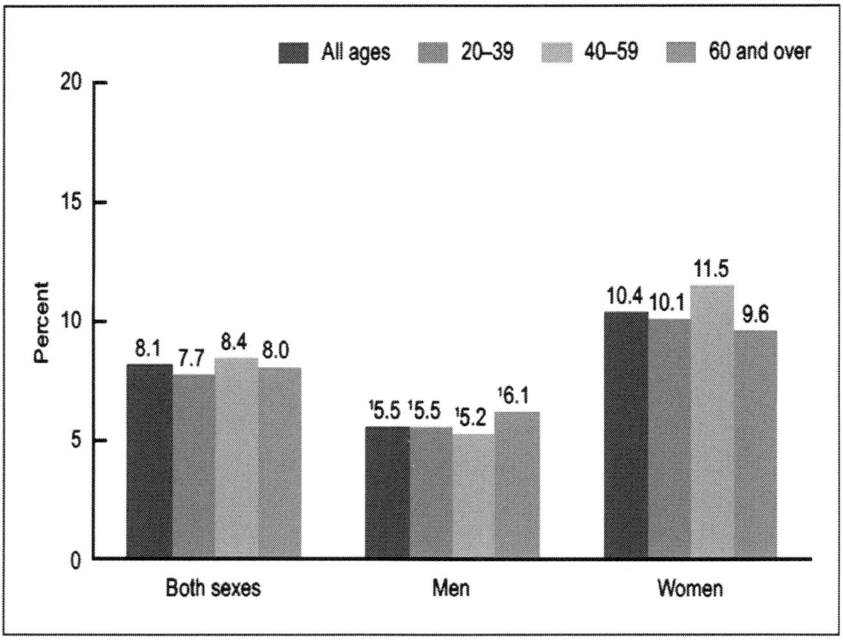

Figure 3.1: Above summarizes the trends of depression from the National Health and Nutrition Examination Survey (Brody et al, 2018: All material appearing in this report is in the public domain and may be reproduced or copied without permission; citation as to source, however, is appreciated).

As the family income levels increased, the levels of depression too showed a dip-in both men and women. 15.8% of individuals below the federal poverty level (FPL) were depressed that dropped to 3.5% when the income was at or more than 400% of the FPL. The highest prevalence was seen in women when the income was lower than that of the FPL with the lowest prevalence in men at or more than 400% of the FPL; 19.8% vs. 2.3%, respectively. Among these depressed individuals, the depression symptoms accounted for difficulty with activities at home, work or society: with moderate or extreme difficulty reported in 30% of the depressed individuals. There was no significant change in depression from 2007–2008 to 2015–2016.

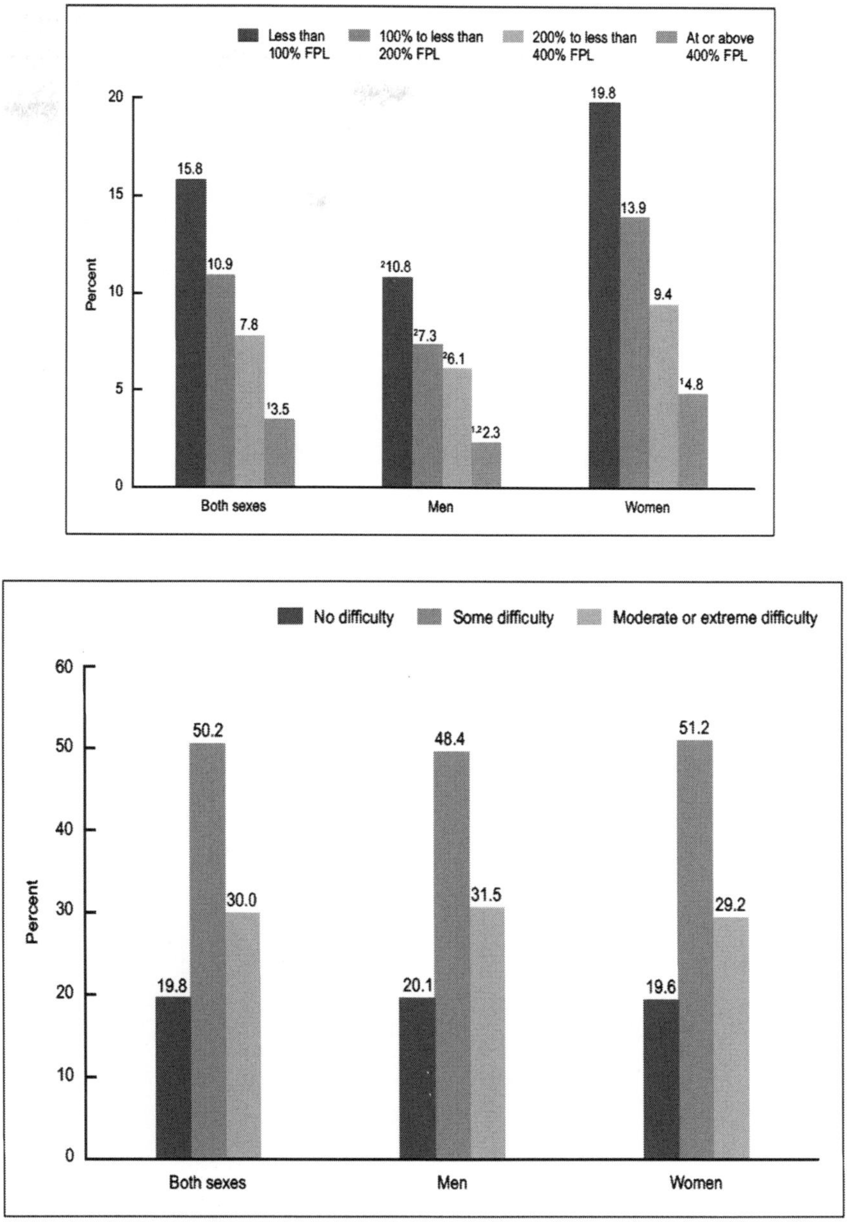

Figure 3.2: Above summarizes the findings of the income and depression from the National Health and Nutrition Examination Survey(Brody et al, 2018: All material appearing in this report is in the public domain and may be reproduced or copied without permission; citation as to source, however, is appreciated).

Chow and team (2019) utilized the 2013 US National Health and Wellness Survey (NHWS) dataset and analyses using the Patient Health Questionnaire-9 (PHQ-9) to assess the cost associated with depression. Of 6997 patients, the initial 6 months showed that the healthcare resource use (HRU), direct and indirect costs and overall impairment of work were elevated when the depression was severe.

3.3 ASSESSING THE BURDEN: WHAT DOES RESEARCH SAY ABOUT DEPRESSION AND DISEASES?

3.3.1 The Sweet Disease, Depression and Costs

Simon and team (2007) evaluated the incremental cost and cost-effectiveness in primary health care clinics of a systematic depression treatment program among outpatients with diabetes.Following the identification of 329 adults with diabetes and current depressive disorder, a 12-month, stepped-care depression treatment program (psychotherapy or anti-depressants) was administered to the patients, the evaluation was done by blinded telephone assessments at 3, 6, 12, and 24 months. It was observed that the outpatient health services costs of patients who received treatments for depression were lower by $314 (mean) against those who followed regular care. In economic terms, the authors state, "When an additional day free of depression is valued at $10, the net economic benefit of the intervention is $952 per patient treated (95% CI, $244 to $1660)".

Hypertension is seen as a comorbidity in patients suffering from diabetes (with close to 3/4th of patients developing the condition)-depression and anxiety are seen in from 15% to 35% in the former condition and 22% to 56% for hypertension. Scientists Wallace and team (2018) explored the economic burden of depression and anxiety in adults with comorbid diabetes and hypertension. Utilizing the 2013 and 2015 Medical Expenditure Panel Survey (MEPS) data, 4560 adults were identified showing both the conditions. Following the division of the samples into 4 groups: no depression/anxiety, depression only, anxiety only, and depression and anxiety-the health-related quality of life (HRQoL) and economic burden were assessed. 13.2% of the 4560 patients reported depression while anxiety was seen in 8.7%, and 7.7% showed both conditions.

Table 3.1: Below displays the physical component summary (PCS) and mental component summary (MCS) scores among adults with comorbid diabetes and hypertension (Wallace et al, 2018: This is an open-access article distributed under the Creative Commons Attribution License, which permits unrestricted use, distribution, and reproduction in any medium, provided the original work is properly cited)

Unadjusted means and SE				
	PCS		MCS	
	Mean (SE)	value	Mean (SE)	value
Depression and anxiety	37.27 (0.96)	<0.001	40.45 (0.94)	<0.001
Depression only	37.02 (0.97)	<0.001	43.76 (0.71)	<0.001
Anxiety only	38.38 (1.04)	0.016	46.96 (0.78)	<0.001
No depression/anxiety (reference group)	40.92 (0.34)		52.8 (0.23)	
Parameter estimates and SE from multivariable OLS regressions[†]				
	PCS		MCS	
	Beta (SE)	value	Beta (SE)	value
Depression and anxiety	−0.68 (0.87)	0.432	−10.35 (0.91)	<0.001
Depression only	−1.37 (0.83)	0.098	−7.67 (0.66)	<0.001
Anxiety only	0.01 (0.80)	0.987	−4.88 (0.74)	<0.001
No depression/anxiety (reference group)				

The annual expenditure for comorbid patients with depression and anxiety for a third party payer exceeded by$9132 (95% CI: $8956–$9304) compared to those who had only diabetes-hypertension. The out-of-pocket (OOP) spending was more for the former: $399 (95% CI: $393–$406) than those with neither depression nor anxiety.

Table 3.3: This is presented below (Wallace et al, 2018: This is an open-access article distributed under the Creative Commons Attribution License, which permits unrestricted use, distribution, and reproduction in any medium, provided the original work is properly cited).

Bivariate association between depression and anxiety categories and high OOP spending burden						
	High burden†		Not high burden			
		Wt row%		Wt row%	value	Sig.
Depression and anxiety	90	34.0	219	66.0	<0.001	
Depression only	158	28.2	403	71.8		
Anxiety only	96	23.5	270	76.5		
No depression/ anxiety	650	18.3	2674	81.7		
Logistic regressions on high OOP spending burden†						
	Unadjusted model			Adjusted model‡		
	OR	95% CI	Sig.	AOR	95% CI	Sig.
Depression and anxiety	2.26	(1.66–3.07)		1.55	(1.06–2.25)	
Depression only	1.72	(1.28–2.32)		1.25	(0.91–1.70)	
Anxiety only	1.34	(1.04–1.74)		1.03	(0.77–1.38)	
No depression/ anxiety	(reference group)					

In a nutshell, comorbid diabetes and hypertension when present along with depression and anxiety account for not only a lowered quality of life but also account for an increased economic burden with an increased third party and out of the pocket spending. This calls for screening for depression at timely intervals to lower the clinical and economic loads on patients who are already not at ease.

(Wallace et al, 2018)

This sets the stage for looking at treating depression associated with diseases to bestow benefits economically when looked at from the point of

health care plans warranting the necessity of screening for and addressing depressive symptoms along with the regular course of treatment.

3.3.2 The Cost of Depression in the Crab

The terms, "you have cancer", are often associated with negative emotions ranging from fear to distress. This makes depression a common outcome of cancer diagnosis. Early studies have pointed out to depression occurring in 38% of patients with cancer of the pancreas to 58% in individuals hospitalized with terminal illnesses (Pan and Sambamoorthi, 2015).

A study published in 2012 evaluated the extent of depressed cancer survivors (adult but non-elderly) and the cost associated with healthcare for the same. A cohort of 11,014 cancer survivors was assembled from data of military health care beneficiaries with the annual health care utilization and costs determined for the year 2009. Of the dataset, the team reported that, "12.6% had a comorbid diagnosis of depression at the time of or after a cancer diagnosis. The highest rates of diagnosed depression occurred in those with cancers of the esophagus, pancreas, ovary, or bronchus, lung, or other respiratory organ; and were associated with female sex, single marital status, and enlisted sponsor rank. Survivors who were diagnosed with depression had significantly higher health care utilization for inpatient and outpatient services, more medication prescriptions, and higher annual costs".

(Jeffery and Linton, 2012)

According to Jayadevappa and team (2012), analyses of the 1995 -2003 Surveillance, Epidemiology and End Results-Medicare linked database for patients who received a diagnosis of prostate cancer between 1995 and 1998 subjected to identification and retrospective follow-up for 1-year pre-diagnosis and up to 8 years post-diagnosis revealed that 8.54% of 50,147 patients newly diagnosed with prostate cancer. As against men who did not have depression, the men who were depressed reported higher healthcare costs as well as more "emergency room visits (odds ratio (OR) = 4.45, 95% CI = 4.13, 4.80), hospitalizations (OR = 3.22, CI = 3.08, 3.37), outpatient visits (OR = 1.71, CI = 1.67, 1.75) and excess risk of death over the course of the follow-up interval (hazard ratio = 2.82, CI = 2.60, 3.06)."

In another 2015-published work, analyses of the medical conditions ofcancer survivors above 21 years of age from the 2006-2009 Medical Expenditure Panel Survey (MEPS) revealed that depression was reported in 14% of individuals (4,766 adult survivors)with cancer. In these individuals expressing both depression and cancer, the average annual health care expenditures accounted to $18,401 against $12,091 in those without depression(Pan and Sambamoorthi, 2015).

Mausbach and team (2018) estimated that the total annual healthcare charges were higher in patients diagnosed with cancer at the University of California San Diego Healthcare System having depression as against individuals with no depression.

From the point of view of a patient and a payer, the healthcare expenditures of elderly cancer patients were assessed by researchers Gu and team (2020). On the basis of the 2007–2013 Cost and Use files of Medicare Current Beneficiary Survey (MCBS)-Medicare sponsored by the Centers for Medicare & Medicaid Services (CMS), identification of patients with breast, lung and prostate cancer (age: 65 years and over)who were newly diagnosed so between 2007 and 2012. On the basis of the questions, "In the last 12 months, how much of the time did you feel sad, blue or depressed? " and "In the last 12 months, have you had 2 weeks or more when you lost interest or pleasure in the things that you usually cared about or enjoyed?", depression was identified in 17.7% or 128 of 710 elderly cancer (breast, lung and prostate) patients.

Table 3.4: The details of the elderly patients are represented below (Gu et al, 2020: Open Access -This article is licensed under a Creative Commons Attribution 4.0 International License, which permits use, sharing, adaptation, distribution and reproduction in any medium or format, as long as you give appropriate credit to the original author(s) and the source, provide a link to the Creative Commons licence, and indicate if changes were made. The images or other third party material in this article are included in the article's Creative Commons licence, unless indicated otherwise in a credit line to the material. If material is not included in the article's Creative Commons licence and your intended use is not permitted by statutory regulation or exceeds the permitted use, you will need to obtain permission directly from the copyright holder. To view a copy of this licence, visit http://creativecommons.org/licenses/by/4.0/. The Creative Commons Public Domain Dedication waiver (http://creativecommons.org/publicdomain/zero/1.0/) applies to the data made available in this article, unless otherwise stated in a credit line to the data)

Characteristics	Without Depression		With Depression	
	N	Wt%	N	Wt%
Total	582	82.3	128	17.7
Predisposing				
Year of cancer diagnosis				
2007–2009	317	83.9	63	16.1
2010–2012	265	80.7	65	19.3
Gender				
Female	234	79.5	56	20.5
Male	348	84.2	72	15.8
Age				
65–74	177	79.4	48	20.6
75 and over	405	84	80	16
Race/ethnicity				
Non-Hispanic white	520	82.9	112	17.1
Other	62	76.8	16	23.2
Enabling				
Marital status				
Married	349	82.8	75	17.2
Other	233	81.2	53	18.8
Educational attainment				
Less than high school	114	77.9	34	22.1
High school	213	85.3	39	14.7
Greater than high school	255	81.8	55	18.2
Poverty status				
LT 200% FPL	435	84.8	79	15.2
GE 200% FPL	147	74.8	47	25.2
Supplemental insurance**				
Private insurance with Rx	237	83.1	51	16.9
Public insurance with Rx	53	64.2	27	35.8
Medical Insurance only	259	86.1	42	13.9

Other	33	81.9	8	18.1
Need				
Cancer site				
Breast	208	80.8	44	19.2
Lung	67	80.6	19	19.4
Prostate	307	83.7	64	16.3
Perceived health status* **				
Excellent/very good/Good	475	86.4	75	13.6
Fair/poor	107	67.3	53	32.7
Functional status limitation* **				
None	426	84.9	78	15.1
≥1	156	64.3	50	35.7
Number of comorbid health conditions **				
None or 1	195	88.9	28	11.1
>1	387	79.1	100	20.9
Personal health practices and use of health services				
Smoking Status				
Current	38	77.9	11	22.1
Past	319	79.9	76	20.1
Never	225	86.6	41	13.4
BMI				
Underweight/ normal	206	80.4	54	19.6
Overweight	258	83.7	49	16.3
Obese/morbid obese	118	82.5	25	17.5
External Environment				
Metro status				
Metropolitan	414	82.5	87	17.5
Non-Metropolitan	168	81.5	41	18.5

***P <.001, **.001 ≤ P <.01, *.01 ≤ P <.05

1. Note: Wt% Weighted percentage, LT less than, GE greater than or equal to, FPL federal poverty level, Rx prescription coverage, BMI body mass index

Patients who had depression displayed conspicuously elevated total healthcare expenditures of $70,918 against $44106 in those with no depression. The expenditure incurred for medical provider services was $25,052 against $16,068 for patients with vs. those without depression. Looking at the payer's point of view, Medicare payments were $48,875 vs. $28,856 for patients with vs. those without depression.

Table 3.5: The details of the expenditures incurred by elderly patients are represented below (Gu et al, 2020: Open Access -This article is licensed under a Creative Commons Attribution 4.0 International License, which permits use, sharing, adaptation, distribution and reproduction in any medium or format, as long as you give appropriate credit to the original author(s) and the source, provide a link to the Creative Commons licence, and indicate if changes were made. The images or other third party material in this article are included in the article's Creative Commons licence, unless indicated otherwise in a credit line to the material. If material is not included in the article's Creative Commons licence and your intended use is not permitted by statutory regulation or exceeds the permitted use, you will need to obtain permission directly from the copyright holder. To view a copy of this licence, visit http://creativecommons.org/licenses/by/4.0/. The Creative Commons Public Domain Dedication waiver (http://creativecommons.org/publicdomain/zero/1.0/) applies to the data made available in this article, unless otherwise stated in a credit line to the data)

Healthcare Expenditures	**Without Depression (N = 582)**			**With Depression (N = 128)**		
	N	**Mean $**	**SE $**	**N**	**Mean $**	**SE $**
Total healthcare expenditures						
Overall* **	582	44,106	2116	128	70,918	5759
By service types						
Medical provider***	582	16,068	934	128	25,052	2609
Hospital outpatient	582	8050	658	128	8006	865
Prescribed medicine	582	7891	485	128	10,188	1242
In Users†						
Inpatient	206	28,743	1890	77	35,712	3785
Other*	430	3559	286	97	8653	2152
By payers						

Medicare***	582	28,856	1716	128	48,875	4150
Out-of-pocket(patient)	582	6511	291	128	9442	1516
Other third-party payers	582	7950	402	128	11,722	2053
Medicare healthcare expenditures						
By service types						
Medical Provider**	582	10,832	700	128	15,566	1545
Hospital outpatient	582	5766	501	128	5949	673
In Users†						
Inpatient	198	25,658	1850	75	31,072	4458
Prescribed medicine	299	5868	624	69	9659	1969
Other**	103	7077	932	48	12,218	1652
Out-of-pocket healthcare expenditures						
By service types						
Medical provider**	582	1903	122	128	3028	348
Prescribed medicine	582	1639	98	128	1667	189
Other	582	2067	158	128	4020	1316
In Users†						
Inpatient	74	3290	911	27	1685	441
Hospital outpatient	359	823	112	80	659	191

***P <.001, **.001≤ P <.01, *.01≤P<.05

1. † Because a large number of patients did not have expenditures in these categories of expenditures, these expenditures were compared only among users (i.e., patients with non-zero expenditures)
2. Note: SE Standard Error

Following adjustment for independent variables, the total healthcare expenditures were 34.5% ($11,454) greater in patients with depression. From the perspective of the payer, the expenditure increased by 32.9% ($1270) while from Medicare's perspective, the expenditure was on the higher side by 43.8% ($8280). Interestingly, this increased expenditure was not associated with chronic nature or self-reported health or education or race or sex.

Table 3.6: The adjusted expenditures of depressed vs. non-depressed elderly cancer patients are below (Gu et al, 2020: Open Access -This article is licensed under a Creative Commons Attribution 4.0 International License, which permits use, sharing, adaptation, distribution and reproduction in any medium or format, as long as you give appropriate credit to the original author(s) and the source, provide a link to the Creative Commons licence, and indicate if changes were made. The images or other third party material in this article are included in the article's Creative Commons licence, unless indicated otherwise in a credit line to the material. If material is not included in the article's Creative Commons licence and your intended use is not permitted by statutory regulation or exceeds the permitted use, you will need to obtain permission directly from the copyright holder. To view a copy of this licence, visit http://creativecommons.org/licenses/by/4.0/. The Creative Commons Public Domain Dedication waiver (http://creativecommons.org/publicdomain/zero/1.0/) applies to the data made available in this article, unless otherwise stated in a credit line to the data).

	AOR [95% CI]	**Coefficient (SE)**	**$ Change [95% CI]**	**% Change [95% CI]**
Overall		0.30 (0.09)**	11,454 [4472,19,729]	34.5 [13.5,59.3]
By service types				
Medical provider		0.38 (0.1)***	8213 [3477,13,998]	45.9 [19.4,78.1]
Hospital outpatient		−0.79 (0.14)	−617 [−2387,1702]	−7.6 [−29.5,21.0]
Prescribed medicine		−0.07 (0.11)	−217 [−819,531]	−6.5 [−24.6,16.0]
Inpatient‡	2.94 [1.82,4.74]***	0.05 (0.11)	1061 [−3036,6137]	5.3 [−15.0,30.0]
Other‡	1.05 [0.65,1.69]	0.41 (0.16)*	405 [69,870]	50.1 [8.5107.4]
By Payers				
Medicare		0.37 (0.1)***	8280 [3570,13,977]	43.8 [18.87,73.91]
Out-of-pocket(patient)		0.28 (0.13)*	1270 [139,2720]	32.9 [3.60,70.40]
Other		0.23 (0.15)	2613 [−546,6826]	26.1 [−5.5,68.2]

	AOR[95% CI]	Coefficient (SE)	$ Change [95% CI]	% Change [95% CI]
Medicare healthcare expenditures				
Medical provider		0.31 (0.1)*	4327 [1425,7856]	36 [11.8,65.4]
Hospital outpatient		−0.02 (0.14)	−97 [1180,1327.9]	−2.1 [−25.6,28.8]
Inpatient‡	2.7 [1.59,4.58]***	0.05 (0.12)	922 [3374,6389]	4.8 [−17.6,33.4]
Prescribed medicine‡	0.88 [0.53,1.46]	−0.07 (0.17)	−76 [−387,363]	−6.7 [33.9,31.7]
Other‡	2.55 [1.59,4.09]*	0.39 (0.17)*	870 [113.21921]	47.2 [6.1104.1]
Out-of-pocket healthcare expenditures				
Medical provider		0.39 (0.16)*	654 [104,1407]	47.1 [7.5101.3]
Prescribed medicine		−0.02 (0.1)	−10 [−94,93]	−2.3 [−20.9,20.7]
Other		0.43 (0.2)*	465 [21,1130]	53 [2.3128.7]
Inpatient‡	1.71 [0.97,3.01]	−0.54 (0.35)	−1025 [−1740,403]	−41.8 [−70.9,16.4]
Hospital outpatient‡	1.05 [0.58,1.92]	−0.26 (0.22)	−342 [−751,296]	−23 [−50.5,19.9]

***P<.001, **.001≤P<.01, *.01≤P<.05

1. ‡ Because a large number of patients did not have expenditures in these categories of expenditures, two-part models, with logistic regressions in the first part and GLMs with gamma distribution and log link in the second part were used to estimate the adjusted effect of depression
2. Note: SE, Standard Error

In a nutshell, the increased healthcare utilization and expenditures from both the point of view of the payer and patient in cancer patients with depression as against those with no depression set the stage for optimal management and addressing of depression in these patients to not only improve the health but also lower the cost incurred. Given that the elderly fall under the vulnerable category, appropriate and integrated systems for treatment are the need of the hour.

3.4 FINAL POINTS

Analyses of 2010 data revealed that US$2.5 trillion was the estimated cost of mental disorders: both direct and indirect. Compared to other major diseases such as cancer or cardiac diseases, indirect costs were more for mental issues- US$1.7 trillion for the latter vs. US$0.8 trillion for the former diseases. A country with highly developed healthcare systems like the EU displayed an estimated cost of €798 billion, with figures expected to rise by a factor of two by 2030! (These costs exclude illicit drug-associated mental issues!)Though the impact of mental issues is at par with that of physical issues, much is spent on the latter despite these grim figures (Trautmann et al, 2016).

Constant awareness and reminders about the impact of mental health can usher in a change with investment in controlling, preventing, and treating mental health challenges like depression while lowering the economic burden on a patient with a "dis"ease.

3.5 REFERENCES

1. Birnbaum HG, Kessler RC, Kelley D, Ben-Hamadi R, Joish VN, Greenberg PE. Employer burden of mild, moderate, and severe major depressive disorder: mental health services utilization and costs, and work performance. Depress Anxiety. 2010;27(1):78-89. doi: 10.1002/da.20580. PMID: 19569060.
2. Greenberg PE, Fournier AA, Sisitsky T, Pike CT, Kessler RC. The economic burden of adults with major depressive disorder in the United States (2005 and 2010). J Clin Psychiatry. 2015 Feb;76(2):155-62. doi: 10.4088/JCP.14m09298. PMID: 25742202.
3. Egede, L. E., Bishu, K. G., Walker, R. J., & Dismuke, C. E. (2016). Impact of diagnosed depression on healthcare costs in adults with and without diabetes: United States, 2004-2011. Journal of affective disorders, 195, 119–126. https://doi.org/10.1016/j.jad.2016.02.011.
4. Brody DJ, Pratt LA, Hughes J. Prevalence of depression among adults aged 20 and over: United States, 2013–2016. NCHS Data Brief, no 303. Hyattsville, MD: National Center for Health Statistics. 2018.
5. Smith, J. P., & Smith, G. C. (2010). Long-term economic costs of psychological problems during childhood. Social science & medicine (1982), 71(1), 110–115.
6. Goodman, A., Joyce, R., & Smith, J. P. (2011). The long shadow cast by childhood physical and mental problems on adult life. Proceedings of the National Academy of Sciences of the United States of America, 108(15), 6032–6037. https://doi.org/10.1073/pnas.1016970108 https://doi.org/10.1016/j.socscimed.2010.02.046
7. McMillan KA, Enns MW, Asmundson GJ, Sareen J. The association between income and distress, mental disorders, and suicidal ideation and attempts: findings from the Collaborative Psychiatric Epidemiology Surveys. J Clin Psychiatry. 2010 Sep;71(9):1168-75. doi: 10.4088/JCP.08m04986gry. Epub 2010 Apr 6. PMID: 20441719.
8. Wing Chow, Michael J. Doane, Jack Sheehan, Larry Alphs, Hoa Le. Economic Burden Among Patients With Major Depressive Disorder: An Analysis of Healthcare Resource Use, Work Productivity, and Direct and Indirect Costs by Depression Severity.
9. Simon GE, Katon WJ, Lin EH, Rutter C, Manning WG, Von Korff M, Ciechanowski P, Ludman EJ, Young BA. Cost-effectiveness of systematic depression treatment among people with diabetes

mellitus. Arch Gen Psychiatry. 2007 Jan;64(1):65-72. doi: 10.1001/archpsyc.64.1.65. PMID: 17199056.

10. Kimberly Wallace, Xiaohui Zhao, RanjitaMisra, Usha Sambamoorthi, "The Humanistic and Economic Burden Associated with Anxiety and Depression among Adults with Comorbid Diabetes and Hypertension", Journal of Diabetes Research, vol. 2018, Article ID 4842520, 9 pages, 2018. https://doi.org/10.1155/2018/4842520.
11. Pan X, Sambamoorthi U. Health care expenditures associated with depression in adults with cancer. J Commun Support Oncol. 2015;13(7):240–7.
12. Jeffery DD, Linton A. The impact of depression as a cancer comorbidity: rates, health care utilization, and associated costs. Commun Oncol. 2012;7(9):216–21.
13. Jayadevappa R, Malkowicz SB, Chhatre S, Johnson JC, Gallo JJ. The burden of depression in prostate cancer. Psychooncology. 2012 Dec;21(12):1338-45. doi: 10.1002/pon.2032. Epub 2011 Aug 12. PMID: 21837637.
14. Mausbach BT, Yeung P, Bos T, Irwin SA. Health care costs of depression in patients diagnosed with cancer. Psycho-Oncology. 2018;27(7):1735–41.
15. Gu, D., Morgan, R.O., Li, R. et al. Association between depression and healthcare expenditures among elderly cancer patients. BMC Psychiatry **20,** 131 (2020). https://doi.org/10.1186/s12888-020-02527-x
16. Trautmann, S., Rehm, J., &Wittchen, H. U. (2016). The economic costs of mental disorders: Do our societies react appropriately to the burden of mental disorders?. EMBO reports, 17(9), 1245–1249. https://doi.org/10.15252/embr.201642951

CHAPTER 4

THE SOCIAL ASPECTS OF DEPRESSION: INSIGHTS

CONTENTS

4.1 LONELINESS, SOCIAL ISOLATION, AND DEPRESSION: RESEARCH AND FACTS

"A lonely man is a lonesome thing, a stone, a bone, a stick, a receptacle for Gilbey's gin, a stooped figure sitting at the edge of a hotel bed, heaving copious sighs like the autumn wind" (Cheever, 1991)

Given the social nature of humans, healthy bonding with humans becomes vital for overall health. Research too backs up the vital involvement of social relationships on health: several factors make their appearance ranging from access to resources to social activities to social support. Empirically, the effect of social isolation or interactions that are considered negative is not encouraging.

For example, Schuster and team (1990) discussed that "regression analyses suggest that negative interactions are more predictive of depressed mood than supportive interactions (specifically with spouse and friends)".

Whisman and Bruce (1999) explored the occurrence of a major depressive episode (MDE) over 12 months on account of marital dissatisfaction. The incidence of MDE was found to be elevated by a factor of 3 due to marital dissatisfaction in a community sample (n=904) of married individuals from the New Haven Epidemiologic Catchment Area program. Close to 30% of the depressive incidences were attributed to dissatisfaction on the marriage front with no influence of depression history or the gender suggestive to be an important factor to depression.

2004-published research involved 240 working individuals of both genders filling the revised UCLA loneliness scale [a 20-point questionnaire rated as Never (1 point) to Often (4 points) to assess the extent of loneliness and social isolation]. The scores of loneliness displayed a dip in married individuals vs. divorced or single participants. Higher loneliness scores displayed a prominent association with low self-esteem, hopelessness, and social isolation. In the case of women, acute mental stress displayed a positive correlation with diastolic blood pressure. The levels of fibrinogen and natural killer cell activity were more in lonely participants. It was concluded that "loneliness is a psychological experience with potentially adverse effects on biological stress processes that may be relevant to health" (Steptoe et al, 2004).

Cacioppo et al (2014) looked at how the risk of broad-based morbidity and mortality are elevated due to a lack of social associations considered as positive. This becomes relevant today given that chronic diseases and their associated treatment costs are on the rise in developed and industrialized countries. The researchers analyzed how much a person feels "isolated or lonely" in today's world deemed to be "social". It was observed that wellbeing (physical and mental) and sleep are adversely impacted by loneliness. This loneliness was also found to boost thesensitivity to social threats. On a positive note, loneliness was found to augment the need of a person to renew social connections.

Holt-Lunstad and Smith (2010) assessed the implications of loneliness in terms of public health.

Databases (n=8) were explored for these aspects followed by subsequent measurement tool to assess systematic reviews (AMSTAR) and the grading of recommendations, assessment, development and evaluations (GRADE) analyses. The authors stated that "*Meta-analyses have identified a significant association between social isolation and loneliness with increased all-cause mortality* and social isolation with cardiovascular disease. Policy makers and health and local government commissioners should consider social isolation and loneliness as important upstream factors impacting on morbidity and mortality due to their effects on cardiovascular and mental health".

A meta-review conducted by scientists Holt-Lunstad and team (2015) involved the net extent of loneliness and social isolation. By employing MEDLINE, CINAHL, PsycINFO, Social Work Abstracts, and Google Scholar, the team assessed how social isolation or loneliness or staying alone influenced mortality in papers published between January 1980 and February 2014. Statistical analyses revealed that the average mortality likelihood increased by 29% for those who were socially isolated; 26% for lonely individuals, and 32% for those persons living alone. The observations displayed consistency for the location in the world, gender, and follow-up time. In cases where social deficits were observed for people below 65 years of age, mortality could be predicted at a higher rate. It was concluded that "Overall, the influence of both objective and subjective social isolation on risk for mortality is comparable with well-established risk factors for mortality".

Such studies have led to the intervention of public governments to address this social isolation and loneliness. An example is 'social prescribing' launched by the UK government in 2018 for tailor-based monitoring of people.According to Prime Minister Theresa May, "Loneliness is a reality for too many people in our society today… it can affect anyone of any age and background…This strategy is only the beginning of delivering a long and far reaching social change in our country - but it is a vital first step in a national mission to end loneliness in our lifetimes".

2020-published data by Menec and team explored the association between health and loneliness. 4 datasets were created as 1) the 'socially isolated and lonely'; 2) the 'only socially isolated'; 3) the 'only lonely'; and, 4) the 'neither socially isolated nor lonely' employing data from the Canadian Longitudinal Study on Aging (CLSA). This data contained 30,079 participants; the Maintaining Contact Questionnaire (MCQ) used for the study was completed by 28,789participants.

Measures	Weighted N	Weighted % or Mean	Standard Error
Age groups			
Ages 45–54	1,563,066	42.0	0.38
Ages 55–64	1,108,198	29.8	0.31
Ages 65–74	638,899	17.2	0.23
Ages 75–85	407,614	11.0	0.18
Sex			
Female	1,869,254	50.3	0.37
Male	1,848,523	49.7	0.37
Education			
Less than postsecondary	1,179,197	31.7	0.34
Postsecondary	2,538,392	68.3	0.34
Household income (yearly)			
< $20,000	161,709	4.4	0.13
$20,000 to < $50,000	655,621	17.6	0.25
> = $50,000	2,694,978	72.5	0.30
Missing response	205,468	5.5	0.16
Functional status			
No functional impairment	3,413,505	92.3	0.18
Mild, moderate, severe, total impairment	286,007	7.7	0.18
Number of chronic diseases	3,744,848	2.94	0.02
Depressive symptom	3,734,419	1.29	0.00
Desire for more social participation			
No	1,886,290	50.5	0.37

Yes	1,849,218	49.5	0.37
Social support			
Tangible support	3,740,050	4.27	0.01
Positive interactions	3,740,399	4.26	0.01
Affection	3,740,263	4.49	0.01
Emotional support	3,740,602	4.23	0.01
Psychological distress			
Low	3,089,408	87.2	0.26
High	455,251	12.8	0.26

	Total Sample	Neither isolated nor lonely	Only isolated	Only lonely	Isolated and lonely
	100	74.7	15.6	6.7	3.0
Covariates					
Ages 45–54	42.0	44.3[a]	35.4[b]	37.1[b]	32.3[b]
Ages 55–64	29.8	29.4	30.4	31.9	31.9
Ages 65–74	17.2	16.5	20.0	16.4	20.3
Ages 75–85	11.0	9.8	14.1	14.6	15.5
Women	50.3	49.6[a]	49.9[a]	60.0[b]	47.9[a]
Men	49.7	50.4	50.1	40.0	52.1
Low education	31.7	30.8[a]	31.2[a]	39.9[b]	40.0[b]
High education	68.3	69.2	68.8	60.1	60.0
< $20,000	4.4	2.1[a]	10.2[b]	7.6[b]	21.7[c]
$20,000 < $50,000	17.6	14.8	25.4	25.3	31.9
> = $50,000	72.5	78.0	58.0	60.4	38.5
No response	5.5	5.1	6.4	6.7	7.9
No functional impairment	92.3	93.7[a]	89.6[b]	86.4[c]	82.2[c]
Mild/moderate/severe/total functional impairment	7.7	6.3	10.4	13.6	17.8
Number chronic conditions	2.94	2.79[a]	3.23[b]	3.58[c]	3.76[c]
Depressive symptom	1.29	1.2[a]	1.26[b]	1.90[c]	1.95[c]
Outcome measures					
No desire for more participation	50.5	52.7[a]	50.6[a]	33.2[b]	33.6[b]
Desire for more participation	49.5	47.3	49.4	66.8	66.4

Tangible support	4.27	4.41[a]	3.93[b]	3.89[b]	3.30[c]
Positive interaction	4.26	4.39[a]	4.04[b]	3.76[c]	3.37[d]
Affection	4.49	4.65[a]	4.16[b]	4.09[b]	3.42[c]
Emotional support	4.23	4.33[a]	4.06[b]	3.78[c]	3.49[d]
Low psychological distress	87.2	89.9[a]	86.9[b]	66.6[c]	62.7[c]
High psychological distress	12.8	10.1	13.1	33.4	37.3

Percentages or means that do not share a superscript are significantly different from each other based on weighted, bivariate least square mean comparisons derived from regression analyses (logistic regressions for dichotomous outcomes, ordinary least squares regressions for continuous outcomes, and ordinal regressions for ordinal level outcomes) using a Bonferroni adjustment (p value of .01/6 = .0017).

Figure 4.1: Above shows **TOP:** the details of the participants; **BOTTOM:** the status in these participants: 15.6% were only socially isolated, 6.7% were only lonely, and 3% were socially isolated and lonely while neither socially isolated nor lonely participants accounted for 74.7%(Menec et al, 2020:This is an open-access article distributed under the terms of the Creative Commons Attribution License, which permits unrestricted use, distribution, and reproduction in any medium, provided the original author and source are credited).

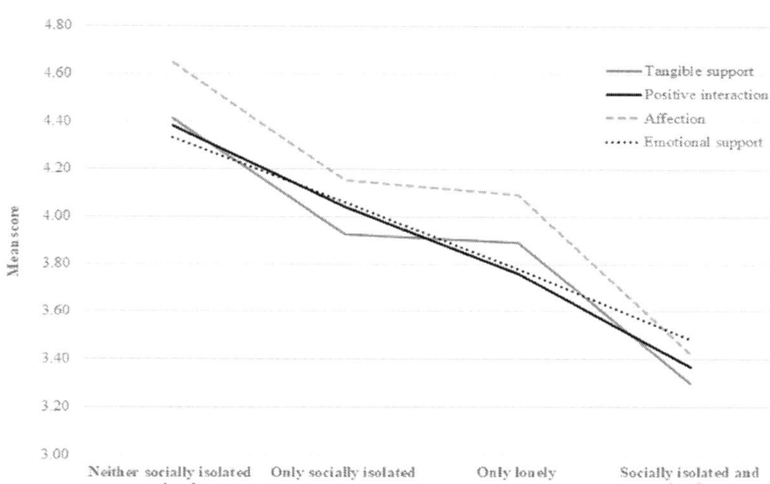

Figure 4.2: The only socially isolated group reported lesser positive interactions and emotional support than that of neither socially isolated nor lonely group. However, the former reported a more positive trend in terms of these aspects than the socially isolated and lonely group (Menec et al, 2020:*This is an open-access article distributed under the terms of the Creative Commons Attri-*

bution License, which permits unrestricted use, distribution, and reproduction in any medium, provided the original author and source are credited).

The number of women was found to be the most in the only lonely group while the highest proportion of low-income individuals was observed in the socially isolated and lonely group. Analyses showed that chronic disease or functional challenges were lower inneither the socially isolated nor the lonely group compared to the socially isolated group. While these aspects were lower for the only lonely group and the socially isolated and lonely group than compared to the socially isolated group, there were no significant variations for the only lonely group and the socially isolated and lonely group.

	Tangible support		Positive interactions		Affection		Emotional support	
	Unadjusted[1]	Adjusted	Unadjusted	Adjusted	Unadjusted	Adjusted	Unadjusted	Adjusted
Social isolation/ loneliness groups								
Neither isolated nor lonely vs. only isolated	0.47 (0.02)	0.39 (0.02)	0.34 (0.01)	0.26 (0.01)	0.48 (0.01)	0.40 (0.01)	0.27 (0.01)	0.20 (0.01)
Neither isolated nor lonely vs. only lonely	0.51 (0.02)	0.34 (0.02)	0.62 (0.02)	0.43 (0.02)	0.54 (0.02)	0.38 (0.02)	0.54 (0.02)	0.37 (0.02)
Neither isolated nor lonely vs. isolated and lonely	1.13 (0.04)	0.87 (0.04)	1.03 (0.03)	0.74 (0.03)	1.24 (0.04)	0.96 (0.04)	0.87 (0.03)	0.60 (0.03)
Only isolated vs. only lonely	0.04 (0.03)	-0.05 (0.03)	**0.28 (0.02)**	**0.17 (0.02**	0.06 (0.03)	-0.02 (0.03)	**0.27 (0.02)**	**0.17 (0.02)**
Only isolated vs. isolated and lonely	0.66 (0.04)	0.48 (0.04	0.69 (0.04)	0.48 (0.03)	0.76 (0.04)	0.56(0.04)	0.59 (0.03)	0.39 (0.03)
Only lonely vs. isolated and lonely	0.62 (0.04)	0.53 (0.04)	0.42 (0.04)	0.31 (0.04)	0.70 (0.05)	0.59 (0.04)	0.33 (0.04)	0.23 (0.04)

1Parameter estimates are shown and, in brackets, standard errors. The analyses are based on Baseline data, unweighted N = 29,644 for tangible support and positive interaction; unweighted N = 29,642 for affection; and unweighted N = 29,645 for emotional support. Parameter estimates are derived from weighted ordinary least squares regression and test differences in least squares means between groups. Adjusted analyses control for: age group, sex, education, household income, functional impairment, chronic conditions, depressive symptom, and province of residence at baseline. Statistical significance was assessed using a Bonferroni adjustment, p value of .01/6 = .0017. Significant results are bolded.

	High psychological distress (vs. low distress)	
Social isolation/loneliness groups	Unadjusted[1]	Adjusted
Neither isolated nor lonely vs. only isolated	-0.30 (0.06)	-0.11 (0.06)
Neither isolated nor lonely vs. only lonely	-1.51 (0.06)	-0.66 (0.08)
Neither isolated nor lonely vs. isolated and lonely	-1.65 (0.08)	-0.65 (0.10)
Only isolated vs. only lonely	-1.21 (0.08)	-0.56 (0.09)
Only isolated vs. isolated and lonely	-1.35 (0.09)	-0.55 (0.11)
Only lonely vs. isolated and lonely	-0.14 (0.10)	0.01 (0.12)

1Parameter estimates are shown and, in brackets, standard errors. Psychological distress is derived from the Maintaining Contact Questionnaire; unweighted N in the analyses = 28,042. Parameter estimates are derived from a weighted logistic regression and test differences in least squares means between groups. Adjusted analyses control for: age group, sex, education, household income, functional impairment, chronic conditions, depressive symptom, and province of residence at baseline. Statistical significance was assessed using a Bonferroni adjustment, p value of .01/6 = .0017. Significant results are bolded.

Figure 4.3: Avove shows the results of the analyses for **TOP:** social support and **BOTTOM:** psychological distress across the datasets(Menec et al, 2020:This is an open-access article distributed under the terms of the Creative Commons Attribution License, which permits unrestricted use, distribution, and reproduction in any medium, provided the original author and source are credited).

The risk was observed for the socially isolated, lonely, or both datasets with the most risk group emerging as the socially isolated and lonely group. It was concluded that though this group may not be significant in terms of number of affected individuals, intervention becomes vital for them given that they face challenges in terms of social support and undergo psychological distress. A positive observation was the motivation in this group to establish and augment social contacts. Points to be noted are the low income and challenges to social contact-the former can be addressed by increasing the affordability of medical treatment while the latter can be addressed using social programs or online social interactions using low-cost options.

The only lonely group displayed risk termed as the second rank with reports of psychological distress and the desire to establish more social contacts despite the presence of a social network underlying the need for "meaningful" associations.

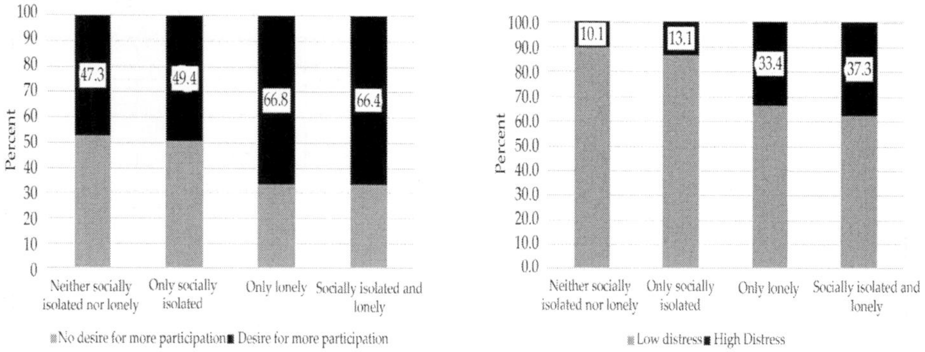

Figure 4.4: Above shows the desire for social connections and the extent of distress in the analyzed population (Menec et al, 2020:This is an open-access article distributed under the terms of the Creative Commons Attribution License, which permits unrestricted use, distribution, and reproduction in any medium, provided the original author and source are credited).

To summarize, loneliness and social isolation are factors that warrant attention and intervention. To quote the authors, "the complex findings suggest that it would be important in future research to gain a more in-depth picture of relationship patterns and social support in different age groups, such as the quality of people's relationships, actual support received, and who provides support"(Menec et al, 2020).

4.2 ASSESSING THE LONGITUDINAL IMPACT OF SOCIETY

"Those who are nurtured best, survive best."

— Louis Cozolino, The Neuroscience of Human Relationships: Attachment And the Developing Social Brain

The aforementioned study presented highlights the vital involvement of mental health and social relationships. While we have been familiar with the concept of "family, friends, networking" since childhood, let us look at what research has to say in this regard.

2000-published research explored the involvement of the relationship with a spouse, relatives, and friendsin Axis I disorders. On the basis of the Ontario Health Survey Mental Health Supplement, analyses of 4933 participants revealed that 6 of 9 Axis I disorders analyzed:generalized anxiety disorder, major depression, panic, and alcohol problems displayed a strong involvement of disagreements with the spouse. This early work thus unearthed the importance of healthy marital associations for mental health (Whisman et al, 2000).

On the basis of the Research Network on Successful Midlife Development (MIDMAC) survey termed as MIDUS (Midlife Development in the United States), several findings emerged that have been given below:

- While the health of women is worse than men, maintenance of health is more in the former with men showing a greater social gradient for ill health.
- A good life according to adults involves the development of relationships with others as the primary factor followed by health and finally family.

(Sutphen, 2004)

Research by Teo and team (2013) explored the role of social relationships in depression both qualitatively and quantitatively. This involved assessing the risk of depression and the involvement of negative/positive relationships, the quality of relationships, and the effect of social isolation.

The national longitudinal cohort study called the Midlife in the United States (MIDUS) served as the sample. The cohort was inclusive of surveys from 4,642 American adults between 25-75 years of age from 1996-1996 and at a follow-up post 10 years. Social support and strain accounted for relationship quality while the reported frequency of social contact and the presence or absence of a partner accounted for social isolation.

Characteristic	n	% (weighted)
Demographic		
Female	2500	52.5
Age category		
25–34 years	881	28.9
35–44 years	1238	27.3
45–54 years	1191	19.9
55–64 years	851	12.3
65–75 years	481	11.6

Ethnicity (n=4,581)		
White	4261	92.8
Black	177	3.7
Other	143	3.4
Socioeconomic		
Education (n=4,634)		
Some grade school to some high school	324	6.5
Graduated high school/GED	1275	26.8
Some college	1406	30.6
Graduated college or higher	1629	36.1
Household income, dollars (n=4,536)		
0–25,000	798	17.9
25,001–50,000	1170	26.2
50,001–100,000	1435	32.0
100,001+	1133	24.0
Health Status		
Very good or excellent overall physical health (n=4,635)	2608	57.6
Very good or excellent overall mental health (n=4,639)	2946	64.6
Major depressive episode	559	12.3
Generalized anxiety disorder	110	2.3
Alcohol misuse (n=4,571)	625	15.5
	Mean (weighted)	**SD (weighted)**
Quality of Social Relationships [a]		
Overall poor quality with spouse/ partner, family, and friends (n=)	1.81	0.36
Social strain with spouse/partner (n=3,538)	2.22	0.60
Social strain with family (n=4,555)	2.10	0.60
Social strain with friends (n=4,600)	1.94	0.50
Lack of social support from spouse/partner (n=3,545)	1.39	0.54
Lack of social support from family (n=4,594)	1.55	0.60
Lack of social support from friends (n=4,609)	1.76	0.65
Social Isolation [b]		

Lack of social contact with family (n=4,574)	3.15	1.49
Lack of social contact with friends (n=4,606)	3.36	1.66
Lack of social contact with neighbors (n=4,624)	3.94	1.57
	n	**% (weighted)**
Unmarried or unpartnered	1142	25.1

[a]Overall poor quality, social strain, and lack of social support were measured on a four-point scale, with a higher score indicating poorer quality, more strain, and less support, respectively.

[b]Lack of social contact was measured on an eight-point scale for family and friends and a six-point scale for neighbors, with a higher score indicating more isolation.

Figure 4.5: Above depicts the baseline characteristics of the 4,642 study participants. The baseline reported a major depressive episode in 12.3% of these participants with more reports of strained social relationships (Teo et al, 2013: This is an open-access article distributed under the terms of the Creative Commons Attribution License, which permits unrestricted use, distribution, and reproduction in any medium, provided the original author and source are credited).

Statistical analyses revealed that the risk of depression was elevated when core relationships were found to be of poor quality, across all the types of social relationships (adjusted odds ratio [AOR], 2.65; 95% CI, 1.86–3.76). The risk of depression was higher in participants reporting strain (AOR, 2.03; 95% CI, 1.49–2.76) and lack of support (AOR, 1.79; 95% CI, 1.36–2.36). In terms of the type of social relationship, the risk of depression was elevated for participants who experienced an overall poor quality of relationship with one's spouse/partner (AOR, 1.47; 95% CI, 1.16–1.87) and family members (AOR, 1.45; 95% CI, 1.10–1.90). Subsequent depression risk did not involve a strong role inthe overall quality of relationships with friends (AOR, 1.21; 95% CI,.84–1.72).

	Participants with spouse/partner at baseline			All participants at baseline		
Component of relationship quality at baseline	**n**	**Adjusted OR[a] (95% CI)**	**Robust SE**	**N**	**Adjusted OR[a] (95% CI)**	**Robust SE**
Overall poor quality	3287	2.60 (1.84–3.69)	0.46	4278	1.94 (1.46–2.59)	0.29
Social strain	3326	1.99 (1.47–2.70)	0.31	4312	1.51 (1.18–1.92)	0.19
Lack of social support	3357	1.79 (1.37–2.35)	0.25	4345	1.46 (1.19–1.80)	0.16

[a]Adjusted for the following variables at baseline: major depression, age, ethnicity, sex, household income, education level, generalized anxiety disorder, alcohol misuse, overall physical health, and overall mental health. Social strain, lack of social support, and overall poor quality in relationships were rated on a four-point scale and each was in a separate multivariable model. For analyses with participants with a spouse/partner, scales of social relationships quality are a composite of relationships with spouse/partner, family, and friends; for analyses of all participants, scales contain just the latter two relationships.

doi:10.1371/journal.pone.0062396.t002

	Participants with spouse/partner at baseline			All participants at baseline		
Type of relationship	n	Adjusted OR[a] (95% CI)	Robust SE	n	Adjusted OR[a] (95% CI)	Robust SE
Spouse/partner	3287	1.43 (1.13–1.82)	0.17	n/a	n/a	n/a
Family	3287	1.47 (1.12–1.93)	0.20	4278	1.54 (1.21–1.95)	0.19
Friends	3287	1.20 (0.84–1.72)	0.22	4278	1.23 (0.92–1.65)	0.18

[a]Adjusted for the following variables at baseline: major depression, age, ethnicity, sex, household income, education level, generalized anxiety disorder, alcohol misuse, overall physical health, and overall mental health. Quality in relationships was rated on a four-point scale. Spouse/partner, family, and friends were included in the same multivariable model for analyses of participants with a spouse/partner and the just the latter two for analyses with all participants.

doi:10.1371/journal.pone.0062396.t003

Figure 4.6: Above documents these findings; **TOP**: The involvement of overall social relationships in the risk of depression; **BOTTOM**: the role of the type of social relationship on the risk of depression (Teo et al, 2013: This is an open-access article distributed under the terms of the Creative Commons Attribution License, which permits unrestricted use, distribution, and reproduction in any medium, provided the original author and source are credited).

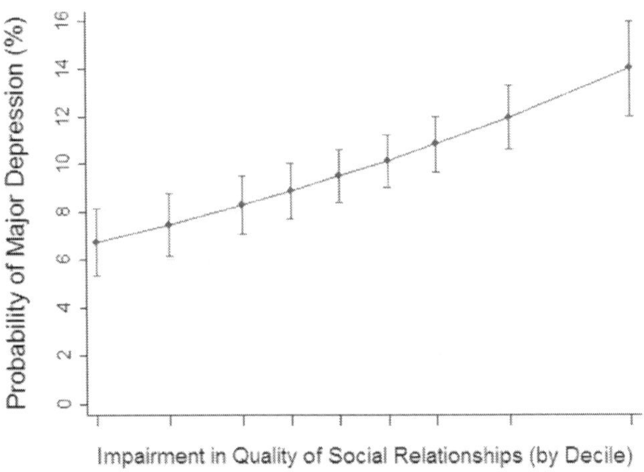

Figure 4.7: The risk of depression striking after 10 years due to the net effect of all the social relationships was assessed using a multivariable logistic regression model. The risk of major depression was 6.7% when the quality of social relationships was the best against 14% in people who reported poor relationships. Interestingly, the risk of depression and the social relationships was found to be "dose-dependent". The data presented is for the general US population with higher impairment of social associations shown as a higher score(Teo et al, 2013: This is an open-access article distributed under the terms of the Creative Commons Attribution License, which permits unrestricted use, distribution, and reproduction in any medium, provided the original author and source are credited).

The authors concluded that one in seven people with poor social relationships and one in fifteen people with good quality associations display a risk of depression. To quote the authors, "Our results thus support the argument that *subjective* components of social relationships are more critical to one's health than *objective* characteristics of one's social network".
(Teo et al, 2013)

This research is suggestive of an overall intervention for depression by augmenting social relationships. This brings in the concept that the subjective nature of relationships of patients should be accounted for by health caregivers to augment the therapeutic response of patients.

4.3 IS AGE INVOLVED IN FEELING SOCIALLY ISOLATED?

*"Our society must make it right and possible for old people not to fear the young or be deserted by them, for the test of a civilization is the way that it cares for its helpless members"-**Pearl S. Buck***

Scientists Adam and team (2004) evaluated the role of lonelinessas a risk factor for depressive symptoms in an elderly dataset (60-98 years of age) living in retirement homes. The factors for predicting depression ranged from age to loss of a loved one to suffering from health challenges to lower participation in social events, and fewer visits from friends. Loneliness emerged as an independent factor for depression given that it accounted for around 8% of the score variances. This set the stage for the authors to propose that loneliness needs to be addressed in senior citizens to circumvent future mental health issues.

Early research in 2004 assessed the importance of social contacts in 113 subjects living in nursing homes; aged 65-101 years. Surveys utilizing the modified Katz Index of ADL, the Revised Social Provisions Scale (SPS), and the Family and Friendship Contacts Scale were performed followed by statistical analyses. Social loneliness was prominently lowered when the elderly subjects had contacts with their children or grandchildren that allowed the authorto suggest the involvement of social contacts while handling the elderly (Drageset, 2004).

2015-published research looked at the risk of all-cause mortality due to loneliness and social isolation over 4 years using data from a longitudinal survey of community-dwelling Singaporean elderly (N=4,522). On the basis of the UCLA three-item loneliness scale, regression was conducted to explore its effect on the risk of mortality. It was observed that all-cause mortality was not accounted for by social networks but the risk of mortality was 44% in sometimes lonely and 39% in mostly lonely. Given this higher risk of mortality, loneliness, and mental health become significant issues that need to be addressed (Chan et al, 2015).

Variable	Overall		Severity of isolation,%				p-value[a]	Loneliness,%		p-value[a]
	N	%	Isolated (n = 504)	High risk (n = 435)	Moderate risk (n = 461)	Low risk (n = 519)		Not lonely (n = 1796)	Lonely (n = 123)	
Age, Mean			59.3	51.4	47.0	46.3	<0.001	51.0	50.8	0.079
21–39	569	29.7	12.7	26.3	38.8	40.8		29.3	35.2	
40–59	696	36.3	35.7	41.8	34.1	34.1		36.7	29.5	
60–74	477	24.9	33.3	23.6	23.0	19.5		25.1	21.3	
≥ 75	177	9.2	18.3	8.3	4.1	5.6		8.9	13.9	
Gender							0.885			0.699
Male	842	43.8	44.2	43.2	45.1	42.8		43.7	45.5	
Female	1,077	56.2	55.8	56.8	54.9	57.2		56.3	54.5	
Marital status							<0.001			<0.001
Single	490	25.6	19.2	21.6	29.5	31.6		24.7	37.7	
Married	1,225	63.8	64.1	68.7	59.9	63.0		65.5	40.2	
Widowed/ Divorced	204	10.6	16.7	9.7	10.6	5.4		9.8	22.1	
Ethnicity							0.010			0.563
Chinese	1,506	78.5	76.2	79.5	81.2	77.5		78.6	77.2	
Malay	151	7.9	11.9	7.6	5.6	6.2		7.7	10.6	
Indian	213	11.1	10.1	9.9	10.8	13.3		11.2	8.9	
Others	49	2.6	1.8	3.0	2.4	3.1		2.5	3.3	
Highest education							<0.001			0.157
No formal education	229	11.9	26.3	10.6	4.6	5.6		11.6	17.9	
Primary	238	12.4	19.4	11.8	10.8	7.5		12.5	11.4	
Secondary (sec)	571	29.8	33.5	33.6	26.2	26.0		30.1	24.4	
Post-sec&above	881	45.9	20.8	44.0	58.4	60.9		45.9	46.3	
Employment status							<0.001			<0.001
Employed	1,222	63.7	50.4	64.3	72.4	68.5		64.7	48.8	
Unemployed	81	4.2	8.7	4.1	2.0	1.9		3.7	12.2	
Inactive	616	32.1	40.9	31.6	25.6	29.6		31.6	39.0	
Living arrangement							<0.001			<0.001
Alone	96	5.0	7.1	5.5	3.7	3.7		5.0	5.7	
With spouse, no child	275	14.3	17.5	13.3	13.2	13.1		14.9	6.5	
With child(ren), no spouse	174	9.1	14.3	7.4	10.4	4.2		8.6	16.3	
With spouse &child(ren)	890	46.4	43.1	51.7	43.7	47.5		47.4	32.5	
With others only	484	25.2	18.1	22.1	29.0	31.5		24.3	39.0	
Money insufficiency		14.4	30.2	12.0	10.8	4.4	<0.001	13.0	35.8	<0.001
Smoking status							<0.001			0.433
Never smoked	1,465	76.3	70.9	74.4	77.4	82.1		76.6	71.5	
Current smoker	240	12.5	14.5	13.4	11.1	11.2		12.4	14.6	
Former smoker	215	11.2	14.7	12.2	11.5	6.7		11.0	13.8	
Alcohol abuse	517	26.9	13.3	30.0	34.7	30.8	<0.001	26.9	27.0	0.981
Number of chronic conditions, Mean			1.6	1.0	0.9	0.7	<0.001[b]	1.0	1.7	<0.001[c]
(SD)			(1.7)	(1.3)	(1.3)	(1.3)		(1.4)	(2.0)	
Diagnosis of depression			5.0	1.6	0.6	0.8	<0.001	1.7	7.4	<0.001

The above numbers reflect weighted column %.
[a]p-value obtained using Chi-square test.
[b]p-value obtained using Kruskal-Wallis test.
[c]p-value obtained using Independent-samples Mann-Whitney U test.

https://doi.org/10.1371/journal.pone.0182145.t001

Figure 4.8: In order to assess the role of*social isolation indicators*: defined as marital status, social connections, and living arrangements on loneliness and hence, depression, researchers Ge and team (2017) utilized a representative health survey in the Central region of Singapore. A sample set of 1942 individuals above 21 years of age was assembled on the basis of the Population Health Index Survey (PHI). A structured questionnaire was employed that addressed health conditions, living conditions, marital status, social connectedness, and the three-item UCLA Loneliness Scale. 5.0% were alone while 69.8% were with a spouse or child(Ge et al, 2017: This is an open-access article distributed under the terms of the Creative Commons Attribution License, which permits unrestricted use, distribution, and reproduction in any medium, provided the original author and source are credited).

Interesting research published in 2013 examined the impact of loneliness on overall mortality and social isolation. On the basis of the 2004-2005 English Longitudinal Study of Ageing and a questionnaire-based assessment of 6500 individuals (more than 52 years of age), social contacts and all-cause mortality were assessed to check whether social isolation causes loneliness to, in turn, adversely impact health. Initial analyses revealed that social isolation and loneliness elevated mortality. Data adjustment further showed that loneliness did not influence mortality conspicuously (hazard ratio 0.92, 95% confidence interval, 0.78-1.09) while social isolation evidently impacted the mortality(hazard ratio 1.26, 95% confidence interval, 1.08-1.48 for the top quintile of isolation). The risk seen due to social isolation did not involve the effect of loneliness. Overall, it was concluded that well-being is adversely impacted by both loneliness and social isolation with the latter factor emerging as a stronger contender to be addressed in terms of tackling mortality (Steptoe et al, 2013).

A feeling of isolation was reported by 26.3%-that was observed in participants who were elder, divorced or widowed, unemployed, inactive, and a history of smoking. Loneliness was reported by 6.4% of the participants: interestingly, the age groups where this was reported were 21–39 years (35.2% vs. 29.3%) and more than 75 years (13.9% vs. 8.9%).

50% of participants who felt lonely were socially isolated as seen by social isolation indicators. 47.4% of the individuals who were lonely were without children or spouse while 32.8% were living with children or spouse. In terms of marital status, 40.2% of the participants who felt lonely were married while 65.5% of the married participants did not feel so

Loneliness and scores of depression were moderately correlated ($r = 0.32, p<0.001$) while the correlation was low for depression scores and social connectedness ($r=0.13\text{-}0.15, p<0.001$). The scores for depression were the highest for those who were socially isolated or those who felt lonely.

		Not lonely (n = 1,796)		Lonely (n = 123)	
		n	%	n	%
Marital status	Single	444	24.7	46	37.7
	Married	1,176	65.5	49	40.2
	Widowed/Divorced	176	9.8	27	22.1
Living arrangement	live alone	89	5.0	7	5.7
	with spouse, no child	267	14.9	8	6.6
	with child(ren), no spouse	154	8.6	20	16.4
	with spouse and child(ren)	850	47.4	40	32.8
	with others only	436	24.3	48	38.5
LSNS-6	Isolated	443	24.7	61	50.0
	High risk	405	22.5	30	24.6
	Moderate risk	443	24.7	18	14.8
	Low risk	506	28.2	13	10.7

The above percentages reflected weighted column %.

https://doi.org/10.1371/journal.pone.0182145.t002

Variables	**Coefficients**		
	Regression 1	**Regression 2**	**Regression 3**
Age	-0.03	-0.05	-0.01
Female (Ref. Male)	0.07**	0.06**	0.06**
Ethnicity (Ref. Chinese)			
Malay	0.03	0.02	0.02
Indian	0.00	0.01	0.01
Others	0.05*	0.05*	0.03
Employment status (Ref. Employed)			
Unemployed	0.15***	0.12***	0.09***
Inactive	0.01	0.00	-0.01
Money insufficiency	0.16***	0.13***	0.10***
Currently smoking (Ref. not smoking)	0.01	0.01	0.01
Alcohol abuse	0.08***	0.08**	0.07**
Previous diagnosis of depression	0.23***	0.22***	0.19***
Number of chronic conditions	0.18***	0.17***	0.14***
Marital status (Ref. Single)			
Married		-0.11	-0.06
Widowed /Divorced		0.01	0.00
Living arrangement (Ref. With spouse & child(ren))			
Alone		-0.07*	-0.05
With spouse, no child		0.01	0.01
With child(ren), no spouse		-0.01	0.00
With others only		-0.04	-0.02
Social connectedness with relatives		-0.06*	-0.04
Social connectedness with friends		-0.11***	-0.07**
Loneliness			0.33***
R^2	0.18	0.20	0.30
ΔR^2	-	0.02	0.10

* p<0.05.
**p<0.01.
***p<0.001.

https://doi.org/10.1371/journal.pone.0182145.t004

Figure 4.9: Below summarizes these results of the effects of loneliness and isolation on depression scoresGe et al, 2017: *This is an open-access article distributed under the terms of the Creative Commons Attribution License, which permits unrestricted use, distribution, and reproduction in any medium, provided the original author and source are credited*).

Importantly, this study highlights the vital involvement of isolation on depressive symptoms. This brings in the role of social connectedness with relatives and friends with a more predominant role attributed to the latter in handling depressive symptoms. Importantly, given that research has focused on the effect of isolation and depression in elders, this study shows that middle-aged and young individuals also show depressive symptoms due to social isolation becomes a factor that needs to be addressed by society.

4.4 FINAL POINTS

Given these research findings, a quote is presented here to highlight the importance of quality relationships across all age groups to circumvent depression:

"Too often we underestimate the power of a touch, a smile, a kind word, a listening ear, an honest compliment, or the smallest act of caring, all of which have the potential to turn a life around" – Leo *Buscaglia*

These studies highlight the importance of good quality social relations to address loneliness and limit the feeling of being socially isolated: both of which are associated with several chronic health issues and even the risk of mortality.

This should be taken into account by health care workers, family, friends, and public health officials while addressing depression.

Every human being is the author of his own health or disease.

— Siddhartha Gautama, The Buddha, 563–483 BC

4.5 REFERENCES

1. Cheever J. Journals of John Cheever. Knopf; New York: 1991.
2. Schuster TL, Kessler RC, Aseltine RH (1990) Supportive Interactions, Negative Interactions, and Depressed Mood. American Journal of Community Psychology 18: 423–438.
3. Whisman MA, Bruce ML (1999) Marital dissatisfaction and incidence of major depressive episode in a community sample. Journal of Abnormal Psychology 108: 674–678.
4. Steptoe A, Owen N, Kunz-Ebrecht SR, Brydon L. Loneliness and neuroendocrine, cardiovascular, and inflammatory stress responses in middle-aged men and women. Psychoneuroendocrinology. 2004 Jun;29(5):593-611. doi: 10.1016/S0306-4530(03)00086-6. PMID: 15041083.
5. Cacioppo, J.T. and Cacioppo, S. (2014), Social Relationships and Health: The Toxic Effects of Perceived Social Isolation. Social and Personality Psychology Compass, 8: 58-72. https://doi.org/10.1111/spc3.12087.
6. Holt-Lunstad J, Smith TB, Layton B. Social relationships and mortality risk: A meta-analytic review. PLOS Med. 2010;7: 1–20.
7. Holt-Lunstad J, Smith T, Baker M, Harris T, Stephenson D. Loneliness and social isolation as risk factors for mortality. PerspectPsychol Sci. 2015;10: 227–237. pmid:25910392.
8. MenecVH, Newall NE, Mackenzie CS, Shooshtari S, Nowicki S (2020) Examining social isolation and loneliness in combination in relation to social support and psychological distress using Canadian Longitudinal Study of Aging (CLSA) data. PLoS ONE 15(3): e0230673. https://doi.org/10.1371/journal.pone.0230673.
9. Whisman, M. A., Sheldon, C. T., & Goering, P. (2000). Psychiatric disorders and dissatisfaction with social relationships: Does type of relationship matter? *Journal of Abnormal Psychology, 109*(4), 803–808. https://doi.org/10.1037/0021-843X.109.4.803.
10. Sutphen S. K. (2004). How Healthy Are We? A National Study of Well-Being at Midlife. *Preventing Chronic Disease*, *1*(3), A12.
11. Teo AR, Choi H, Valenstein M (2013) Social Relationships and Depression: Ten-Year Follow-Up from a Nationally Representative Study. PLoS ONE 8(4): e62396. https://doi.org/10.1371/journal.pone.0062396.

12. Drageset J. The importance of activities of daily living and social contact for loneliness: a survey among residents in nursing homes. Scand J Caring Sci. 2004 Mar;18(1):65-71. doi: 10.1111/j.0283-9318.2003.00251.x. PMID: 15005665.
13. Adams KB, Sanders S, Auth EA. Loneliness and depression in independent living retirement communities: risk and resilience factors. Aging Ment Health. 2004;8: 475–485. pmid:15724829.
14. Chan A, Raman P, Ma S, Malhotra R. Loneliness and all-cause mortality in community-dwelling elderly Singaporeans. Demogr Res. 2015;S15: 1361–1382.
15. Steptoe A, Shankar A, Demakakos P, Wardle J. Social isolation, loneliness, and all-cause mortality in older men and women. ProcNatlAcadSci U S A. 2013 Apr 9;110(15):5797-801. doi: 10.1073/pnas.1219686110. Epub 2013 Mar 25. PMID: 23530191; PMCID: PMC3625264.
16. Ge L, Yap CW, Ong R, HengBH (2017) Social isolation, loneliness and their relationships with depressive symptoms: A population-based study. PLoS ONE 12(8): e0182145. https://doi.org/10.1371/journal.pone.0182145

CHAPTER 5

DEPRESSION: LIGHT AT THE END OF THE TUNNEL

CONTENTS

5.1 GUIDELINES FOR CONVENTIONAL TREATMENT

As per the American Psychiatric Association guidelines for the treatment of patients with major depressive disorder, 'the ultimate recommendation regarding a particular clinical procedure or treatment plan must be made by the psychiatrist in light of the clinical data, the psychiatric evaluation, and the diagnostic and treatment options available. Such recommendations should incorporate the patient's personal and socio-cultural preferences and values in order to enhance the therapeutic alliance, adherence to treatment, and treatment outcomes'.

According to Fava (2014), the use of antidepressant drugs emerges as vital and lifesaving when the appropriate indications are sanctioned.

Whilemajor depressive disorder is currently addressed using antidepressants, their effectiveness as displayed in literature either as a group or individually remains debatable. A team of scientists funded by the National Institute for Health Research Oxford Health Biomedical Research Centre and the Japan Society for the Promotion of Science compared and ranked antidepressants for treating adults. This becomes significant as the use of antidepressants supersedes psychological interventions. Based on trials of 21 antidepressants used major depressive disorder diagnosed according to standard operationalized criteria, clinical trials were selected by searching across Cochrane Central Register of Controlled Trials, CINAHL, Embase, LILACS database, MEDLINE, MEDLINE In-Process, PsycINFO, the websites of regulatory agencies, and international registers. This study (PROSPERO number: CRD42012002291) based on 522 trials with 116477 participants entailing the use of the drugs: agomelatine, bupropion, citalopram, desvenlafaxine, duloxetine, escitalopram, fluoxetine, fluvoxamine, levomilnacipran, milnacipran, mirtazapine, paroxetine, reboxetine, sertraline, venlafaxine, vilazodone, and vortioxetine (second-generation antidepressants with approval by the regulatory agencies in the USA, Europe, or Japan), amitriptyline and clomipramine (listed in the WHO Model List of Essential Medicines), along withtrazodone and nefazodone (known for their tolerability and effects) vs. placebo or another individual drug.

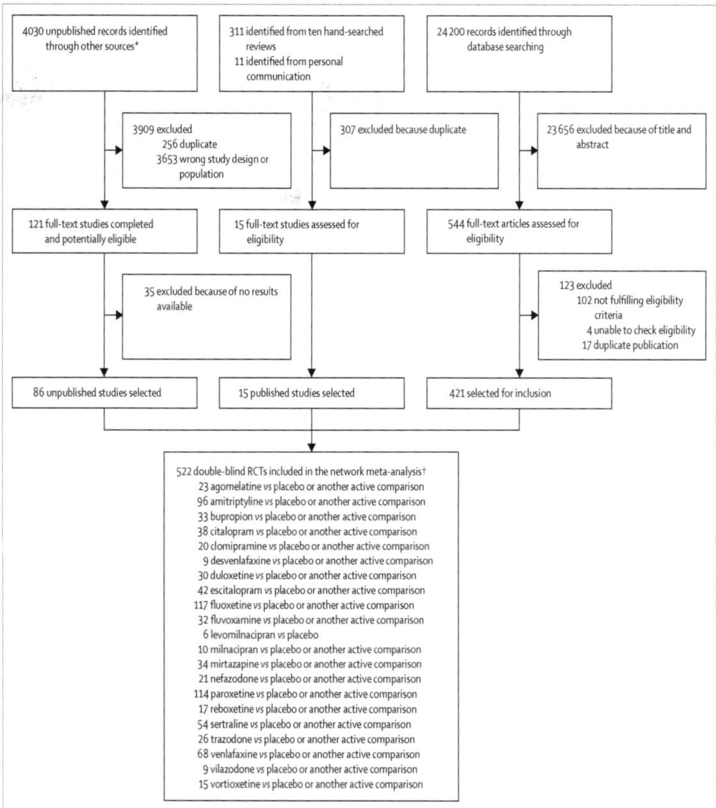

Figure 5.1: Above shows the system followed to compare various antidepressants in the network meta-analysis (Cipriani et al, 2018: This is an open-access article distributed under the terms of the Creative Commons CC-BY license, which permits unrestricted use, distribution, and reproduction in any medium, provided the original work is properly cited).

The response rate (efficacy) and treatment discontinuity (acceptability) were the primary outcomes assessed along with the summary odds ratios (ORs). All the drugs assessed displayed a higher efficacy than the placebo. In terms of acceptability, clomipramine was lower than that of placebo while agomelatine (OR 0·84, 95% CrI 0·72–0·97) and fluoxetine (0·88, 0·80–0·96) had lesser dropouts than the placebo. The efficacies of agomelatine, amitriptyline, escitalopram, mirtazapine, paroxetine, venlafaxine, and vortioxetine were more than the other drugs assessed while the least efficacy was displayed by fluoxetine, fluvoxamine, reboxetine, and trazodone in terms of head-to-head studies.The tolerability was more for agomelatine, citalopram, escitalopram, fluoxetine, sertraline, and vortioxetine while high drop-out rates were seen for amitriptyline, clomipramine, duloxetine, fluvoxamine, reboxetine, trazodone, and venlafaxine.

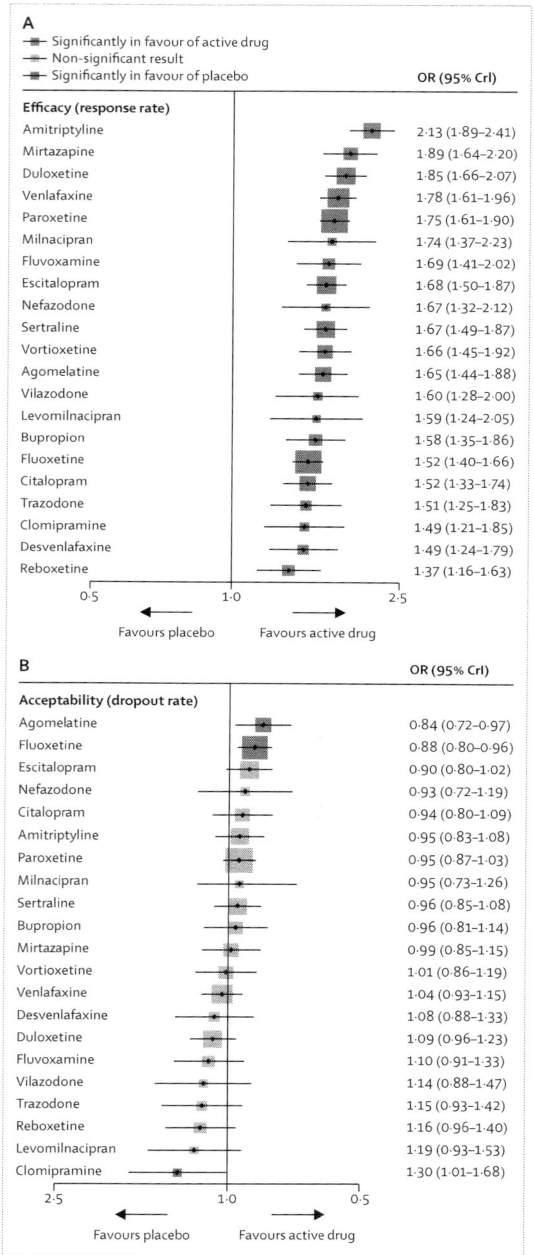

Efficacy (response rate) ■ Comparison Acceptability (dropout rate)

Agom	0·72* (0·55–0·92)	0·80* (0·54–1·15)	0·89* (0·66–1·19)	0·57* (0·42–0·77)	0·62† (0·47–0·82)	0·97* (0·74–1·27)	0·85† (0·68–1·05)	0·69† (0·51–0·97)	0·79* (0·58–1·09)	0·81* (0·61–1·05)	0·70* (0·44–1·14)	0·81* (0·65–1·00)	0·53* (0·36–0·80)	0·86* (0·66–1·13)	0·69* (0·48–0·98)	0·74† (0·58–0·92)	1·24† (0·71–2·19)
0·96* (0·76–1·24)	Amit	1·10‡ (0·78–1·58)	1·23* (0·94–1·64)	0·79† (0·60–1·05)	0·87† (0·66–1·15)	1·35* (1·05–1·74)	1·18† (0·99–1·42)	0·97† (0·74–1·24)	1·10† (0·84–1·45)	1·12* (0·89–1·42)	0·98‡ (0·62–1·55)	1·12† (0·95–1·34)	0·74† (0·51–1·10)	1·20* (0·97–1·47)	0·96‡ (0·70–1·31)	1·02† (0·83–1·26)	1·72† (1·00–3·05)
0·87† (0·59–1·30)	0·91‡ (0·62–1·31)	Bupr	1·11‡ (0·76–1·67)	0·71† (0·49–1·07)	0·78‡ (0·53–1·18)	1·23* (0·84–1·80)	1·07‡ (0·76–1·50)	0·87‡ (0·59–1·30)	1·00‡ (0·66–1·49)	1·01† (0·70–1·47)	0·89‡ (0·51–1·54)	1·02‡ (0·73–1·43)	0·67† (0·42–1·08)	1·08‡ (0·75–1·56)	0·87‡ (0·57–1·30)	0·92‡ (0·66–1·30)	1·55† (0·85–2·94)
1·13* (0·88–1·47)	1·18* (0·93–1·49)	1·30† (0·88–1·93)	Cita	0·64† (0·47–0·87)	0·70* (0·51–0·95)	1·09* (0·85–1·42)	0·96* (0·76–1·21)	0·78* (0·57–1·06)	0·89* (0·64–1·23)	0·91† (0·68–1·21)	0·79‡ (0·49–1·32)	0·91* (0·71–1·17)	0·60† (0·41–0·87)	0·97‡ (0·74–1·25)	0·77* (0·53–1·13)	0·83† (0·64–1·07)	1·40† (0·78–2·48)
1·20* (0·91–1·59)	1·24† (0·98–1·58)	1·37† (0·93–2·04)	1·06* (0·82–1·38)	Clom	1·10† (0·80–1·51)	1·71* (1·27–2·29)	1·49† (1·16–1·90)	1·22‡ (0·88–1·67)	1·40† (1·00–1·92)	1·41* (1·05–1·91)	1·24‡ (0·76–2·00)	1·42† (1·12–1·79)	0·94‡ (0·62–1·41)	1·51† (1·15–1·96)	1·21† (0·83–1·73)	1·29† (0·99–1·67)	2·20† (1·22–3·90)
1·06* (0·82–1·37)	1·10† (0·84–1·42)	1·21† (0·81–1·81)	0·93* (0·71–1·22)	0·88† (0·66–1·18)	Dulo	1·56* (1·19–2·01)	1·37* (1·06–1·73)	1·12* (0·80–1·53)	1·28† (0·91–1·75)	1·30* (0·96–1·72)	1·13‡ (0·69–1·83)	1·30* (1·02–1·63)	0·86‡ (0·57–1·29)	1·38† (1·04–1·80)	1·10† (0·76–1·59)	1·18‡ (0·92–1·49)	1·99† (1·13–3·52)
0·90* (0·71–1·14)	0·93* (0·74–1·17)	1·03† (0·70–1·51)	0·79* (0·65–0·97)	0·75* (0·58–0·97)	0·85* (0·67–1·08)	Esci	0·87* (0·70–1·09)	0·71* (0·53–0·96)	0·81* (0·60–1·11)	0·83* (0·63–1·08)	0·72† (0·45–1·18)	0·83* (0·67–1·03)	0·55* (0·37–0·81)	0·88* (0·69–1·12)	0·70* (0·49–1·00)	0·75* (0·60–0·94)	1·27‡ (0·73–2·25)
1·20* (0·99–1·48)	1·25† (1·06–1·48)	1·38† (0·97–1·97)	1·06* (0·87–1·29)	1·00‡ (0·81–1·24)	1·14* (0·91–1·44)	1·34* (1·11–1·61)	Fluo	0·82* (0·64–1·04)	0·94* (0·72–1·20)	0·95* (0·77–1·16)	0·83‡ (0·54–1·30)	0·95* (0·83–1·09)	0·63† (0·44–0·90)	1·01† (0·84–1·21)	0·81* (0·60–1·09)	0·87† (0·74–1·01)	1·46‡ (0·85–2·53)
1·20* (0·91–1·61)	1·25† (0·99–1·59)	1·38† (0·93–2·07)	1·06* (0·82–1·39)	1·01‡ (0·76–1·32)	1·14† (0·85–1·54)	1·34* (1·03–1·75)	1·00* (0·80–1·25)	Fluv	1·14† (0·84–1·56)	1·16* (0·89–1·52)	1·01‡ (0·62–1·71)	1·16* (0·90–1·49)	0·77‡ (0·51–1·17)	1·23* (0·94–1·63)	0·99‡ (0·69–1·42)	1·06* (0·80–1·38)	1·78‡ (1·00–3·24)
1·07* (0·80–1·44)	1·11† (0·86–1·43)	1·23† (0·81–1·85)	0·94‡ (0·71–1·26)	0·89† (0·67–1·19)	1·01‡ (0·74–1·38)	1·19* (0·90–1·58)	0·89* (0·70–1·13)	0·89† (0·67–1·17)	Miln	1·02† (0·75–1·37)	0·88‡ (0·54–1·44)	1·02‡ (0·80–1·31)	0·67† (0·45–1·03)	1·08* (0·82–1·44)	0·86* (0·60–1·25)	0·93* (0·71–1·22)	1·56‡ (0·89–2·84)
0·93* (0·72–1·21)	0·97* (0·77–1·21)	1·07† (0·73–1·57)	0·82* (0·65–1·05)	0·78* (0·60–1·01)	0·88* (0·67–1·16)	1·04* (0·82–1·32)	0·78* (0·64–0·94)	0·78* (0·60–0·99)	0·87* (0·66–1·15)	Mirt	0·87‡ (0·55–1·41)	1·00* (0·82–1·23)	0·66* (0·45–0·99)	1·06* (0·84–1·35)	0·85* (0·62–1·18)	0·91* (0·73–1·13)	1·53‡ (0·89–2·72)
1·15† (0·76–1·76)	1·19† (0·80–1·78)	1·32‡ (0·80–2·20)	1·01‡ (0·67–1·54)	0·96‡ (0·63–1·45)	1·09‡ (0·71–1·68)	1·28* (0·86–1·94)	0·96‡ (0·66–1·40)	0·95‡ (0·63–1·46)	1·07‡ (0·70–1·67)	1·23* (0·82–1·86)	Nefa	1·15‡ (0·74–1·78)	0·75‡ (0·43–1·32)	1·23† (0·77–1·90)	0·98‡ (0·57–1·64)	1·04‡ (0·66–1·65)	1·76‡ (0·90–3·56)
1·01* (0·82–1·24)	1·05† (0·89–1·23)	1·16† (0·81–1·64)	0·89* (0·72–1·09)	0·84† (0·68–1·03)	0·95† (0·76–1·19)	1·12* (0·93–1·35)	0·84* (0·73–0·95)	0·84* (0·67–1·04)	0·94† (0·75–1·18)	1·08* (0·89–1·30)	0·88‡ (0·60–1·27)	Paro	0·66† (0·46–0·94)	1·06* (0·88–1·28)	0·85† (0·63–1·15)	0·91* (0·77–1·07)	1·53† (0·90–2·66)
1·44* (1·02–2·04)	1·50† (1·07–2·07)	1·65† (1·05–2·60)	1·27† (0·92–1·75)	1·20† (0·84–1·70)	1·36† (0·95–1·95)	1·60* (1·14–2·23)	1·20† (0·88–1·62)	1·20† (0·83–1·71)	1·35† (0·92–1·95)	1·54* (1·09–2·17)	1·25‡ (0·77–2·01)	1·43† (1·05–1·94)	Rebo	1·61† (1·09–2·34)	1·29† (0·81–2·01)	1·38† (0·94–1·99)	2·32† (1·24–4·41)
1·07* (0·85–1·37)	1·11* (0·92–1·35)	1·23† (0·85–1·79)	0·95† (0·76–1·18)	0·90† (0·71–1·13)	1·02‡ (0·79–1·32)	1·20* (0·97–1·48)	0·89‡ (0·76–1·05)	0·89† (0·70–1·13)	1·00† (0·77–1·30)	1·15* (0·93–1·43)	0·93‡ (0·63–1·37)	1·07* (0·90–1·26)	0·75† (0·54–1·04)	Sert	0·80* (0·58–1·11)	0·86* (0·70–1·05)	1·45‡ (0·84–2·54)
1·36* (0·99–1·87)	1·41† (1·06–1·86)	1·56† (1·04–2·31)	1·20* (0·88–1·63)	1·13† (0·83–1·54)	1·28† (0·92–1·79)	1·51* (1·12–2·04)	1·13† (0·87–1·46)	1·13‡ (0·82–1·55)	1·27* (0·91–1·76)	1·45* (1·09–1·94)	1·18‡ (0·75–1·84)	1·35* (1·04–1·75)	0·94‡ (0·64–1·39)	1·26† (0·95–1·67)	Traz	1·07‡ (0·77–1·47)	1·80‡ (0·98–3·38)
1·01* (0·82–1·26)	1·05† (0·87–1·27)	1·16† (0·82–1·65)	0·90† (0·72–1·10)	0·85† (0·67–1·06)	0·96† (0·77–1·21)	1·13* (0·93–1·37)	0·84† (0·73–0·97)	0·84* (0·66–1·07)	0·95* (0·73–1·23)	1·09* (0·89–1·33)	0·88‡ (0·59–1·30)	1·01† (0·86–1·17)	0·70† (0·51–0·97)	0·94* (0·78–1·13)	0·75† (0·57–0·98)	Venl	1·69† (1·01–2·86)
0·73‡ (0·42–1·26)	0·76‡ (0·44–1·29)	0·83‡ (0·45–1·54)	0·64† (0·37–1·11)	0·61† (0·35–1·05)	0·69† (0·40–1·20)	0·81‡ (0·47–1·39)	0·60† (0·36–1·02)	0·60† (0·34–1·05)	0·68† (0·39–1·20)	0·78‡ (0·45–1·34)	0·63† (0·33–1·19)	0·72† (0·43–1·22)	0·51† (0·28–0·92)	0·68† (0·39–1·16)	0·54† (0·30–0·95)	0·72† (0·43–1·19)	Vort

Figure 5.2: Above summarizes the results of the various antidepressants in the network meta-analysis: **TOP**: Meta-analyses of antidepressants vs. placebo: A: Efficacy and B: Acceptability; **BOTTOM**: Acceptability of 21 antidepressants in head-to-head studies presented in alphabetical order. Comparison of ORs (95% CrI) in the column-defining treatment vs. the row-defining treatment; OR=odds ratio. CrI=credible interval. Agom=agomelatine. Amit=amitriptyline. Bupr=bupropion. Cita=citalopram. Clom=clomipramine. Dulo=duloxetine. Esci=escitalopram. Fluo=fluoxetine. Fluv=fluvoxamine. Miln=milnacipran. Mirt=mirtazapine. Nefa=nefazodone. Paro=paroxetine. Rebo=reboxetine. Sert=sertraline. Traz=trazodone. Venl=venlafaxine. Vort=vortioxetine. *Moderate quality of evidence. †Low quality of evidence. ‡Very low quality of evidence. (Cipriani et al, 2018: This is an open-access article distributed under the terms of the Creative Commons CC-BY license, which permits unrestricted use, distribution, and reproduction in any medium, provided the original work is properly cited).

This analysis described by the researchers as "the most comprehensive currently available evidence base to guide the initial choice about pharmacological treatment for acute major depressive disorder in adults", presents data concerning practice to provide a snapshot of antidepressants to patients, doctors, guideline developers, and policy makers.

(Cipriani et al, 2018)

Author's note: The drugs mentioned in this work are only to be taken under medical supervision; this should not be considered as a prescription or treatment guide.

5.2 GUIDELINES FOR USING ANTIDEPRESSANTS

Depression can be targeted using **clinical practice guidelines (CPGs)** that have been formulated on the basis of scientific evidence. The recommendations are inclusive of but not limited to assiduously check for suicidal or self-harm tendencies, screening for depression in populations, efficient therapy, and active monitoring. While the majority of guidelines are on the basis of severity of depression, the recommendations can become specific in the case of comorbidities. For mild depression in its initial stages, the treatment is inclusive of guided self-help, cognitive-behavioral therapy, psychotherapy, and physical activity. For moderate depression, pharmacotherapy or psychotherapy is recommended and for severe depression, both these approaches are followed (Heddaeus et al, 2018).

The extent of adherence to guidelines was assessed in a stepped and collaborative care model (SCM) part of "Health Network Depression" aimed at optimization of diagnoses for prompt screening and efficient care in the Hamburg Network for Mental Health. Researchers Heddaeus and team (2018) assessed the selection of treatment in 569 patients who received care under SCM. The adherence to guidelines was found to be the highest for mild depression followed by moderate depression and finally severe depression. For mild depression,the most selected and frequent approach was bibliotherapy followed by internet-based self-help. Outpatient psychotherapy ranked first followed by pharmacotherapy for moderately depressed individuals. Nearly 60% of severely depressed patients were advised psycho- and pharmacotherapy with only 18.6% however receiving this treatment. In a nutshell, 68% of all the cases were adherent to guidelines while initialization of this treatment was seen in 47% of the cases.

Treatment	Level of depression severity						total (N = 569)	
	mild (n = 85)		moderate (n = 296)		severe (n = 182)			
	TS	TI	TS	TI	TS	TI	TS	TI
active monitoring/GP consultation	**15.3**	**11.8**	5.1	11.5	2.7	12.1	5.8	11.6
bibliotherapy	**40.0**	**38.8**	7.4	7.4	1.6	2.2	10.4	10.4
internet-based self-help	**20.0**	**18.8**	7.1	7.8	0.5	1.6	6.9	7.4
telephone-based psychotherapy	**10.6**	**7.1**	**4.1**	**2.7**	1.1	2.2	4	3.3
psychotherapy	**2.4**	**2.4**	**40.2**	**26.7**	21.4	22.5	28.3	21.6
pharmacotherapy	1.2	2.4	**14.9**	**15.9**	5.5	25.3	9.7	16.9
combined psycho- & pharmaco-therapy	2.4	2.4	6.1	6.8	**53.3**	**18.1**	20.6	9.8
other combination	**2.4**	**5.9**	**8.4**	**8.8**	3.7	5.5	6	7.2
inpatient treatment	0.0	0.0	1	0.0	**6**	**0.5**	2.5	0.2
no intervention	0.0	1.2	0.0	2.0	0.0	0.0	0.0	1.2
missing treatment decision	5.9	-	5.7	-	3.8	0.0	6	0.0
dropout	0.0	9.4	0.0	10.5	0.0	9.9	0.0	10.4
Proportion of guideline adherence	**91%**	**85%**	**68%**	**54%**	**59%**	**19%**	**68%**	**47%**

Bold numbers mark guideline-adherent treatment selection and initialization for the specific severity degree.

https://doi.org/10.1371/journal.pone.0208882.t002

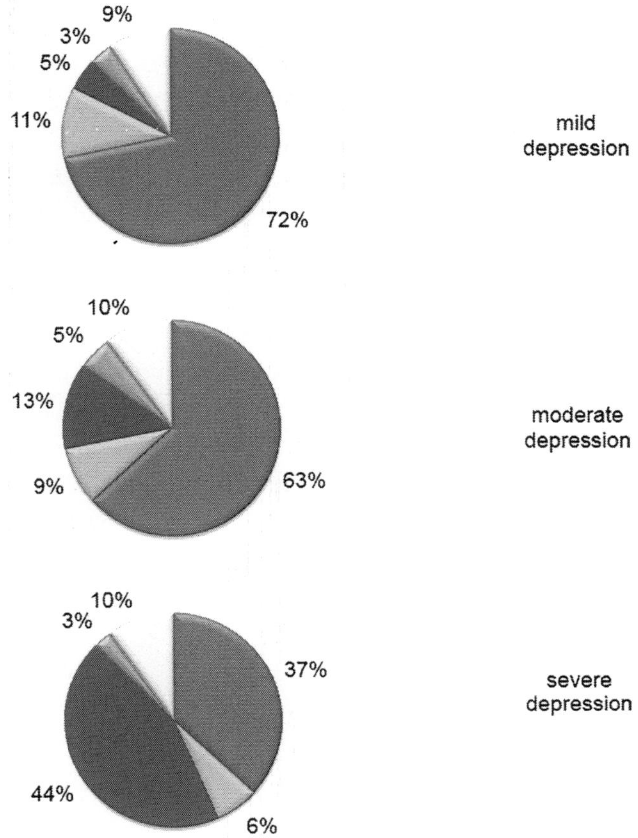

Figure 5.3: Above displays the extent of adherence and initialization of treat-

ments to address depression in a stepped and collaborative care model (SCM). **TOP**: "TS" shows the percentage of treatments selected by patients in the treatment decision process. "TI" is representative of the extent of initialized treatment; **BOTTOM**: Treatment implementation: The decision implemented is in green: treatment decision implemented; orange: treatment implemented more intensive than the selected treatment; lower intensity or no treatment in red: missing data in grey and dropout inwhite (Heddaeus et al, 2018: This is an open-access article distributed under the terms of the Creative Commons Attribution License, which permits unrestricted use, distribution, and reproduction in any medium, provided the original author and source are credited).

Overall, it was concluded that the selection and initialization of treatment are based on the severity of depression while care for mild depression can benefit from SCM application (Heddaeus et al, 2018).

Treatment in patients can be augmented by meticulously designing clinical practice guidelines (CPGs) based on scientific evidence. In order to facilitate this, researchers Gabriel and team (2020) reported a thorough search for CPGs (2011-2016) across MEDLINE, Embase, the Cochrane Library, and CPG-specific databases. The quality of the CPGs was evaluated employing the second version of the Appraisal of Guidelines for Research and Evaluation (AGREE II). From 947 citations, 27 CPGs were subjected to selection from which 6 met the AGREE II Domain 3 score pertaining to the rigor of developing the CPG of ≥80%.

For mild depression, non-pharmacological approaches have been described in a few CPGs. For moderate and severe depression, both pharmacological and psychotherapy have been advocated in CPGs. According to the Canadian Network for Mood and Anxiety Treatments (CANMAT), the first-line treatment entailed the use of agomelatine, mianserin, and milnacipran. Other drugs recommended were mirtazapine (4 CPGs) and amitriptyline, fluoxetine, sertraline, and mirtazapine (Columbian CPG).

Clinical practice guideline/Country/ Development organization	Main characteristics
Guía Clínica AUGE[3] [21] / Chile/ Ministry of Health)	This CPG[1] was prepared in 2013, after two previous versions developed in 2007 and 2009. It has a very well-organized format and flowcharts that allow a clear understanding of recommendations. There are 11 recommendations for pharmacological treatment. Among them, some are intended for specific populations that were not included in this study (e.g., pregnant women and adolescents). This CPG has a methodological manual that recommends the use of a GRADE[2] evidence classification system. However, it makes clear that a simplified system was used to classify the evidence, since the transition to the GRADE methodology is a gradual process that requires specific skills.
Guía de Práctica Clínica [22] / Colombia/ Ministry of Health	This CPG, prepared in 2013, used the adaptation of earlier versions of NICE[4] and CANMAT[5]. It is quite extensive, and its recommendations are organized into clinical questions.
Depression in adults [23] / England / NICE	This CPG was first published in 2009 and was updated in 2016. The full version of the document is quite extensive. It uses GRADE but does not contain the level of evidence or the strength of recommendation explicitly in the recommendations. In 2018, NICE updated this CPG exclusively to make recommendations regarding the use of valproate in pregnancy. These were not included in the current study.
Depression, adult in primary care [24] / USA / ICSI	This CPG comprises many documents. One of them contains the main recommendations clearly highlighted according to their level of evidence and strength of recommendation. Some recommendations are scattered in the text, and do not contain the level of evidence or strength of recommendation.
Practice guideline for the treatment of patients with major depressive disorder [26] / USA / APA	This is the third version of this CPG. APA reaffirmed this guideline in October 2015. It contains several detailed recommendations mainly related to subtypes of depression comorbidities and other clinical conditions associated with depression. It has a simplified evidence classification scale and a manual containing information about its elaboration and development.
Canadian Network for Mood and Anxiety Treatments 2016 [25] / CANADA / CANMAT	This CPG was created in 2001, followed by a subsequent update in 2009. It is currently in its third version. It comprises 21 questions followed by their respective answers. Researchers chose not to follow AGREE II for its development. The developers understood the wide international use of GRADE but opted instead for a self-classification evidence scale. This CPG also contains a separate document in which its development methods are described.

[1]CPG = clinical practice guideline;

[2]GRADE = The Grading of Recommendations Assessment, Development and Evaluation;

[3]AUGE = Aseguramiento Universal de las Garantias Explícitas;

[4]NICE = National Institute for Health and Care Excellence;

[5]CANMAT = Canadian Network for Mood and Anxiety Treatments;

[6]ICSI = Institute for Clinical Systems Improvement;

[7]APA = American Psychiatry Association.

https://doi.org/10.1371/journal.pone.0231700.t001

Figure 5.4: Above displays the clinical practice guidelines selected for analyses (Gabriel et al, 2020: This is an open-access article distributed under the terms of the Creative Commons Attribution License, which permits unrestricted use, distribution, and reproduction in any medium, provided the original author and source are credited).

Table 5.1: Presents the indications and treatment recommendations in clinical practice guidelines (Gabriel et al, 2020: This is an open-access article distributed under the terms of the Creative Commons Attribution License, which permits unrestricted use, distribution, and reproduction in any medium, provided the original author and source are credited)

Recommendations	**Clinical practice guidelines**	**Evidence classification scale**	**Strength of recommendation**	**Level of evidence**	
RISK OF SUICIDE					
Patients on antidepressants and presenting as a suicide risk should be monitored regularly.	[4]	GRADE	Strong	–	
	[5]	GRADE	–	–	
The risk of overdose toxicity is higher in patients using tricyclic antidepressants. NICE, 2016 [5]: except lofrepramine	[4]	GRADE	Strong	–	
	[5]	GRADE	–	–	
Overdose toxicity can occur with all antidepressants but is more with dual-action antidepressants.	[4]	GRADE	Strong	–	
Venlafaxine (a dual-action antidepressant) is associated with a higher risk of death by overdose.	[5]	GRADE	–	–	
The choice of the type of intervention must be influenced by the response to the previous treatment for depression.	[5]	GRADE	–	–	

Pharmacotherapy combined with psychotherapy is a way to increase the intensity of treatment for patients who are at risk of suicide.	[6]	MODIFIED GRADE	Recommended with moderate clinical safety	_	
The need for hospitalization should be considered in patients at risk of suicide.	[6]	MODIFIED GRADE	Recommended with significant clinical safety	_	
ACUTE PHASE TREATMENT					
GOAL					
Reach remission of the depressive episode and return to the patient's base level	[6]	MODIFIED GRADE	Recommended with significant clinical safety	_	
INDICATION FOR PHARMACOLOGICAL TREATMENT					
Antidepressants should not be used routinely to treat mild depression.	[7]	MODIFIED GRADE	Highly recommended, based on good-quality studies	Systematic reviews, meta-analyses, randomized trials, and health technology assessment reports	
	[5]	GRADE	_	_	
	[4]	GRADE	Strong	_	
	[8]	OTHER	_	_	

Antidepressants can be used to treat patients with persistent mild depression or who have a history of moderate/serious depression.* *[5] specifies that this is depression that lasts for at least 2 years.	[4]	GRADE	Strong	–	
	[5]	GRADE	–	–	
Antidepressants are indicated for the treatment of severe depression.	[6]	MODIFIED GRADE	Recommended with significant clinical safety	–	
	[4]	GRADE	Strong	–	
Antidepressants should be used in serious depression unless electroconvulsive therapy is planned.	[6]	MODIFIED GRADE	Recommended with significant clinical safety	–	
Options for treatment are pharmacotherapy, psychotherapy, a combination of pharmacotherapy and psychotherapy, electroconvulsive therapy, transcranial magnetic stimulation, and phototherapy.	[6]	MODIFIED GRADE	–	–	
When the reaction to loss is too extensive or there is functional impairment, psychotherapy, or antidepressants should be used.	[6]	MODIFIED GRADE	Recommended with significant clinical safety	–	

Antidepressants can be a treatment choice for mild-to-moderate depression.	[6]	MODIFIED GRADE	Recom-mended with sig-nificant clinical safety	–	
Antidepressants can be a treatment choice for moderate depres-sion.	[4]	GRADE	Strong	–	
	[6]	MODIFIED GRADE	Recom-mended with sig-nificant clinical safety	–	
INDICATION OF A COMBINATION PHARMACOTHERAPY—PSYCHOTHERAPY					
Consider combin-ing antidepressant and psychotherapy whenever possible	[9]	GRADE	Weak	Moderate	
If combined treat-ment is not possible (pharmacotherapy + psychotherapy), ini-tiate treatment with pharmacotherapy in serious depression.	[9]	GRADE	Weak	Moderate	
For patients with moderate depression consider a combina-tion of antidepres-sants and psycho-therapy.* *[5] suggests cog-nitive-behavioural therapy or interper-sonal psychotherapy	[4]	GRADE	Strong	–	
	[5]	GRADE	Weak	–	
	[6]	MODIFIED GRADE	Recom-mended with sig-nificant clinical safety	–	

For patients with serious depression, use psychotherapy in conjunction with antidepressants.	[7]	MODIFIED GRADE	Highly recommended, based on good-quality studies	Systematic reviews, meta-analyses, randomized trials, and health technology assessment reports	
	[5]	GRADE	–	–	
	[4]	GRADE	Strong	–	
	[6]	MODIFIED GRADE	Recommended with significant clinical safety	–	
For patients with mild depression who have psychosocial or interpersonal issues, intrapsychic conflicts, or axis II disorders, a combination of antidepressants and psychotherapy can be useful.	[6]	MODIFIED GRADE	Recommended with moderate clinical safety	–	
CHOICE OF MEDICATION					

Selective serotonin reuptake inhibitors are recommended as first-line options for antidepressant treatment.	[7]	MODIFIED GRADE	Highly recommended, based on good-quality studies	Systematic reviews, meta-analyses, randomized trials, and health technology assessment reports	
	[5]	GRADE	_	_	
	[4]	GRADE	Strong	_	
	[9]	GRADE	_	_	
	[6]	MODIFIED GRADE	Recommended with significant clinical safety	_	
	[8]	OTHER	_	Level 1	
Tricyclic antidepressants are not recommended as first-line drugs owing to the associated adverse events.	[7]	MODIFIED GRADE	Highly recommended, based on good-quality studies	Systematic reviews, meta-analyses, randomized trials, and health technology assessment reports	
	[9]	GRADE	_	_	

Amitriptyline (a tricyclic antidepressant) can be considered as a first-line drug option. Exception: elderly patients who present contraindications to tricyclic antidepressants.	[4]	GRADE	Strong	–	
For patients with contraindications for the use of tricyclic antidepressants, consider fluoxetine, sertraline, or mirtazapine as first-line options.	[4]	GRADE	Strong	–	
Bupropiona can be considered a first-line drug	[9]	GRADE	–	–	
	[8]	OTHER	–	Level 1	
	[6]	MODIFIED GRADE	Recommended with significant clinical safety	–	
Venlafaxine, duloxetine, desvenlafaxine, and vortioxetine can be considered as first-line drugs.	[9]	GRADE	–	–	
	[8]	OTHER	–	Level 1	
Mirtazapine can be considered a first-line drug option.	[4]	GRADE	Strong	–	
	[9]	GRADE	–	–	
	[8]	OTHER	–	Level 1	
	[6]	MODIFIED GRADE	Recommended with significant clinical safety	–	

Levomilnacipran and vilazodone can be considered as first-line drug options.	[9]	GRADE	–	–	
Agomelatine, citalopram, escitalopram, fluoxetine, fluvoxamine, mianserin, milnacipran, paroxetine, and sertraline	[8]	OTHER	–	Level 1	
CONTINUATION PHASE					
The choice of treatment to reduce the risk of relapse (continuation phase) and recurrence should be discussed with the patient after considering prior treatment, discontinuation symptoms, and patient preferences.	[5]	GRADE	–	–	
Pharmacological treatment should be maintained* for at least 6 months after remission of symptoms. *stipulates a period of 6–12 months [4,7]. *mentions 6–9 months [8].	[7]	MODIFIED GRADE	Recommended with moderate clinical safety	Systematic reviews, meta-analyses, randomized trials, and health technology assessment reports	
	[5]	GRADE	–	–	
	[4]	GRADE	Strong	–	
	[8]	OTHER	–	–	

Pharmacological treatment should extend from 4 to 9 months after remission of symptoms in patients who have undergone successful treatment with antidepressants.	[6]	MODIFIED GRADE	Recommended with significant clinical safety	–	
Explain to the patient that continuing the use of medication after remission considerably reduces the risk of relapse and that antidepressants are not associated with addiction.	[5]	GRADE	–	–	
The same dosage of tricyclic antidepressant, although low, can be continued for patients who had a clear response, maintaining careful monitoring.	[5]	GRADE	–	–	
The same dosage used during the acute phase should generally be used in the continuation phase.	[6]	MODIFIED GRADE	Recommended with moderate clinical safety	–	

People who had several episodes of depression and who responded to treatment with an antidepressant combined with augmentation agents should continue this combination after remission if they find that the adverse events are tolerable and acceptable.	[5]	GRADE	–	–	
For patients who received electro-convulsive therapy, the continuation treatment should be lithium and nortriptyline.	[6]	MODIFIED GRADE	Recommended with significant clinical safety	–	
MAINTENANCE PHASE					
For continued treatment with antidepressants, consider the number* of prior episodes of depression. *[7] specifies that three or more depressive episodes are necessary.	[7]	MODIFIED GRADE	Highly recommended, based on good-quality	Systematic reviews, meta-analyses, randomized trials, and health technology assessment reports	
	[5]	GRADE	–	–	
	[4]	GRADE	Weak	–	
	[6]	MODIFIED GRADE	Recommended with significant clinical safety	–	
	[8]	OTHER	–	–	

For continued treatment with antidepressants, consider the presence of residual symptoms, concomitant physical health issues, and psychosocial difficulties.	[4]	GRADE	Weak	–	
	[5]	GRADE	–	–	
	[6]	MODIFIED GRADE	Recommended with significant clinical safety	–	
	[8]	OTHER	–	Level 3 e Level 4	
The severity of depressive symptoms should also be considered while making decisions regarding the maintenance phase.	[4]	GRADE	Weak	–	
	[5]	GRADE	–	–	
	[8]	MODIFIED GRADE	Recommended with moderate clinical safety	–	
For people with depression who are at a significant risk of relapse or have a history of recurrent depression, explain the treatment options to reduce recurrence, including continuous medication, augmentation of the medication or psychological treatment (cognitive-behavioural therapy).	[5]	GRADE	–	–	
Continue with antidepressants for 2 years if the consequences of relapses can be serious, such as suicide attempts and inability to work.	[5]	GRADE	–	–	

We should extend the treatment for 2 years or more in the presence of risk factors.	[4]	GRADE	Weak	–	
	[8]	OTHER	–	Level 3 Level 4	
	[5]	GRADE	–	–	
Maintain the dosage of the drug that was effective if the adverse events are tolerable.	[5]	GRADE	–	–	
During the maintenance phase, an antidepressant drug that produced remission of symptoms during the acute phase and persistence of remission during the continuation phase should be administered in full therapeutic dosage.	[6]	MODIFIED GRADE	Recommended with moderate clinical safety	–	
When deciding whether to continue the maintenance treatment beyond 2 years, consider the age.	[5]	GRADE	–	–	
A family history of mood disorders and an early age increase the possibility of recurrent episodes and support the need for maintenance treatment.	[6]	MODIFIED GRADE	Recommended with moderate clinical safety	–	
The frequency of depressive symptoms and persistence contribute to decisions regarding the maintenance phase.	[6]	MODIFIED GRADE	Recommended with moderate clinical safety	–	

The patient's preference, the type of treatment received, and the adverse events during the continuation phase should all be considered while making maintenance phase decisions.	[6]	MODIFIED GRADE	Recommended with moderate clinical safety	–	
For patients with chronic depression and concomitant psychiatric treatment, indefinite maintenance treatment is necessary.	[6]	MODIFIED GRADE	Recommended with significant clinical safety	–	
DISCONTINUATION OF TREATMENT					
Before discontinuing the treatment, patients should be informed regarding the risk of relapse.	[6]	MODIFIED GRADE	Recommended with significant clinical safety	–	
Discontinue the antidepressant gradually*to avoid discontinuation symptoms. *NICE, 2016 [5] mentions 4 weeks.	[4]	GRADE	Strong	–	
	[5]	GRADE	–	–	
	[6]	MODIFIED GRADE	Recommended with significant clinical safety	–	
Paroxetine and venlafaxine have a shorter half-life and require a more gradual dosage reduction to minimize discontinuation symptoms.	[4]	GRADE	Strong	–	
	[5]	GRADE	–	–	

Inform the patients that discontinuation symptoms can occur with interruption of treatment or in the event of forgetting dosages.	[6]	MODIFIED GRADE	Recom-mended with sig-nificant clinical safety	–	
	[5]	GRADE	–	–	
When discontinua-tion symptoms occur take necessary action as follows: - If mild, monitor and reassure the patient - If severe, consider reintroducing the antidepressant at the dosage that was effective (or another antidepressant of the same class with a longer half-life) and gradually reduce the dosage while moni-toring the symptoms.	[5]	GRADE	–	–	
If serious discon-tinuation symptoms occur, an antidepres-sant with a longer half-life can be administered to minimize these ef-fects.	[6]	MODIFIED GRADE	Recom-mended with moderate clinical safety	–	
	[5]	GRADE	–	–	
If there is no im-provement in the first month, treat-ment (psychotherapy and/or pharmaco-therapy) should be discontinued.	[6]	MODIFIED GRADE	Recom-mended with sig-nificant clinical safety	–	

In a nutshell, the CPGs analyzed advocated the use of serotonin selective reuptake inhibitors (SSRIs) as the first-line strategy to target depression followed by antipsychotics for patients who were not responsive to SSRIs. A further common point was adjusting the dosage based on the patient's response. While some CPGs offered recommendations based on the depression sub-type, carefully scouring through a CPG before treatment has been advised (Gabriel et al, 2020).

Author's note: The drugs mentioned in this work are only to be taken under medical supervision; this should not be considered as a prescription or treatment guide.

5.3 CAUTION IS NEEDED

Antidepressants have earned the reputation of being the most prescribed medication in young adults, according to reports from the US, Canada and UK (Coupland et al, 2018). According to Moore and Mattison (2017), analyses of the 2013 Medical Expenditure Panel Survey revealed that 16.7% of 242 million US adults had a psychiatric drug prescription: the drugs were 2.0% for antidepressants; 8.3% for hypnotics, anxiolytics, and sedatives, and; 1.6% for antipsychotics. Overall, the researchers showed that psychiatric drugs were consumed by 1 in 6 US adults with 84.3% reporting long-term usage. Interestingly, the use was more with age, in women, and ethnicity.

An analysis of adverse effects of antidepressants was reported by Coupland and team (2018) based on a cohort study of 238,963 patients (20–64 years old). Analyses for risk of falls, fractures, bleeding in the upper gastrointestinal, road traffic accidents, adverse drug reactions and all-cause mortality were done. The rates of falls and fracture rates were elevated with all the classes of antidepressants tested. For tricyclic and related antidepressants, adverse drug reactions and all-cause mortality were on the higher side compared to selective serotonin reuptake inhibitors. Bleeding in the upper gastrointestinal system was observed for trazodone. While selective serotonin reuptake inhibitors were associated with more incidences of fracture, the adverse drug reactions were lower compared to other antidepressants.

Table 5.2: Below displays the 5-year follow-up adjusted hazard ratios of falls, fracture, upper gastrointestinal bleed, adverse drug reaction, road traffic crash, and all-cause mortality in the use of various classes of antidepressants in (A) ages 20 to 44 years and (B) 45 to 64 years (Coupland et al, 2018: Open Access: This article is distributed under the terms of the Creative Commons Attribution 4.0 International License (http://creativecommons.org/licenses/by/4.0/), which permits unrestricted use, distribution, and reproduction in any medium, provided you give appropriate credit to the original author(s) and the source, provide a link to the Creative Commons license, and indicate if changes were made. The Creative Commons Public Domain Dedication waiver (http://creativecommons.org/publicdomain/zero/1.0/) applies to the data made available in this article, unless otherwise stated)

	A. **Ages 20 to 44 years**					B. **Ages 45 to 64 years**				
Antidepressant class	No of events*	Person years*	Adjusted hazard ratio†	95% CI	p	No of events*	Person years*	Adjusted hazard ratio†	95% CI	P
FALLS										
No current use	1373	384,886	1.00			1120	174,047	1.00		
TCAs	116	22,035	1.33	(1.08 to 1.62)	0.006	182	18,225	1.40	(1.19 to 1.63)	<0.001
SSRIs	701	142,227	1.40	(1.28 to 1.54)	<0.001	793	79,586	1.59	(1.44 to 1.75)	<0.001
Other antidepressants	80	16,974	1.28	(1.03 to 1.60)	0.027	109	10,704	1.58	(1.29 to 1.92)	<0.001
Combined antidepressants	13	2,298	1.35	(0.77 to 2.38)	0.29	24	1,823	1.82	(1.22 to 2.73)	0.004
FRACTURE										

No current use	1658	349,837	1.00			1101	160,336	1.00		
TCAs	109	20,204	1.10	(0.90 to 1.34)	0.37	96	17,105	0.76	(0.61 to 0.94)	0.012
SSRIs	812	130,683	1.36	(1.24 to 1.49)	<0.001	654	73,909	1.25	(1.13 to 1.38)	<0.001
Other antidepressants	122	15,331	1.50	(1.25 to 1.80)	<0.001	73	9,834	1.04	(0.82 to 1.32)	0.73
Combined antide-pressants	9	2,086	0.80	(0.42 to 1.56)	0.52	20	1,672	1.62	(1.03 to 2.55)	0.038
UPPER GI BLEED										
No current use	362	391,625	1.00			226	179,029	1.00		
TCAs	34	22,468	1.32	(0.94 to 1.84)	0.11	45	18,827	1.49	(1.05 to 2.11)	0.024
SSRIs	189	144,496	1.32	(1.09 to 1.59)	0.004	111	81,840	0.97	(0.78 to 1.21)	0.81
Other antidepressants	33	17,169	1.66	(1.15 to 2.39)	0.007	17	10,933	0.98	(0.58 to 1.67)	0.95
Combined antide-pressants	8	2,342	2.89	(1.44 to 5.81)	0.003	4	1,903	1.40	(0.51 to 3.82)	0.51
ROAD TRAFFIC CRASH										
No current use	1890	373,475	1.00			453	173,440	1.00		
TCAs	114	21,531	1.03	(0.83 to 1.29)	0.76	40	18,398	0.79	(0.56 to 1.13)	0.20

SSRIs	704	138,656	1.02	(0.93 to 1.12)	0.72	225	79,482	1.06	(0.89 to 1.26)	0.49
Other antidepressants	85	16,521	1.02	(0.81 to 1.28)	0.89	35	10,700	1.27	(0.91 to 1.78)	0.16
Combined antide-pressants	12	2,226	1.06	(0.61 to 1.85)	0.83	6	1,837	1.23	(0.55 to 2.77)	0.62
ADVERSE DRUG REACTION										
No current use	243	392,177	1.00			189	179,224	1.00		
TCAs	58	22,536	2.96	(2.15 to 4.07)	<0.001	64	18,844	2.37	(1.74 to 3.22)	<0.001
SSRIs	249	144,668	2.10	(1.71 to 2.59)	<0.001	147	81,909	1.34	(1.07 to 1.69)	0.011
Other antidepressants	47	17,217	3.83	(2.67 to 5.50)	<0.001	23	10,980	1.78	(1.13 to 2.79)	0.012
Combined antide-pressants	8	2,350	4.92	(2.44 to 9.92)	<0.001	3	1,907	1.36	(0.43 to 4.29)	0.60
ALL-CAUSE MOR-TALITY										
No current use	460	394,758	1.00			1081	180,864	1.00		

TCAs	55	22,743	1.80	(1.35 to 2.41)	<0.001	271	19,064	1.96	(1.68 to 2.28)	<0.001
SSRIs	206	145,580	1.29	(1.07 to 1.55)	0.007	784	82,654	1.43	(1.30 to 1.58)	<0.001
Other antidepressants	43	17,378	1.62	(1.19 to 2.23)	0.003	153	11,109	1.81	(1.50 to 2.18)	<0.001
Combined antide-pressants	10	2,374	2.35	(1.24 to 4.44)	0.009	31	1,926	2.15	(1.50 to 3.08)	<0.001

SSRIs=selective serotonin reuptake inhibitors; TCAs=tricyclic and related antidepressants.

* Based on numbers in adjusted analysis

† Adjusted for age, sex, year of diagnosis of depression, severity of depression, deprivation, smoking status, alcohol intake, ethnic group (white/not recorded or non-white), coronary heart disease, diabetes, hypertension, cancer, epilepsy/seizures, hypothyroidism, osteoarthritis, asthma/chronic obstructive airways disease, stroke/TIA, rheumatoid arthritis, osteoporosis, liver disease, renal disease, obsessive-compulsive disorder, statins, NSAIDS, aspirin, antihypertensive drugs, anticonvulsants, hypnotics/anxiolytics, oral contraceptives, hormone replacement therapy, antipsychotics, bisphosphonates, anticoagulants. Fracture outcome also adjusted for falls.

Though an observational study, the researchers state that given these risks shown with the use of antidepressants, the keyword here is a caution: to tread with care when treating depression.

5.4 HERBS FOR DEPRESSION

Looking at alternative cures to target depression, the use of molecules obtained from plants termed phytochemicals has been explored for several conditions such as depression. Few examples of research reporting the use of herbs or their chemicals are presented below:

A 6-week double-blind, placebo-controlled and randomized trial looked at the effect of Crocus sativus L. (saffron) on 40 adults diagnosed with the Diagnostic and Statistical Manual of Mental Disorders, fourth edition. Following 6 weeks of saffron petal capsules, the scores on the Hamilton Depression Rating Scale were markedly better against the placebo (d.f.=1, F=16.87, p<0.001) thereby showing the suitability of saffron petals to target mild-to-moderate depression (Mishiri et al, 2006).

A cross-sectional study in 1058 Japanese participants (70 or more years old) revealed that 34.1% of the individuals were mildly depressed while 20.2% were severely depressed. The symptoms of depression were lowered with green tea consumption (Niu et al, 2009).

In a pilot study, 28 women considered high-risk postpartumin a Women's center of a large Indianapolis hospital were administered rose otto and Lavandula angustifolia (Lavender) oil as aromatherapy or inhalation. Analyses of the Edinburgh Postnatal Depression Scale (EPDS) and Generalized Anxiety Disorder Scale (GAD-7) scores revealed that there was an evident score improvement in those who took aromatherapy vs. controls. The authors stated, "the pilot study indicates positive findings with minimal risk for the use of aromatherapy as a complementary therapy in both anxiety and depression scales with the postpartum woman" (Conrad et al, 2012).

Other herbs that have been employed to target depression are presented in the tables below:

Table 5.3: Displays phytochemicals that target depression (Gihyun Lee and Hyunsu Bae, 2017: This is an open-access article distributed under the Creative Commons Attribution License, which permits unrestricted use, distribution, and reproduction in any medium, provided the original work is properly cited)

Phytochemical	Dose	Study design	Effects and mechanisms
Carvacrol	12.5–50 mg/kg	Oral administration in mice	Induce antidepressant effects that seem to be dependent on an interaction with the dopaminergic brain pathways
	12.5 mg/kg	Oral administration in rats	Raise 5-HT and dopamine ranges in the hippocampus and prefrontal cortex Influence neuronal activity through modulation of neurotransmitters
Curcumin	1.25–10 mg/kg	Oral administration in rats	Reduce immobility time in the forced swimming test Reverse bilateral olfactory bulbectomy-induced hyperactivity in the open field and deficits in step-down passive avoidance
	20–40 mg/kg	Intraperitoneal injection in mice	Enhance 5-HT level
			Restore biochemical and behavioral changes induced by the chronic stress Reverse the decreased immobility period and MAO activity induced chronic stress
	10 mg/kg	Oral administration in mice	Reduce duration of immobility in forced swimming test May be associated with 5-H and 5-H subtypes
	10–20 mg/kg	Oral administration in rats	Attenuate the stress-induced hippocampus 5-H mRNA
Ferulic acid	100–250 mg/kg	Oral administration in mice	Attenuate stress-induced behavior Increase CREB phosphorylation and brain-derived neurotropic factor mRNA level in the hippocampus

L-Theanine	1–20 mg/kg	Oral administration in mice	Reduce immobility time in the forced swimming test and tail suspension test without ambulation in the open field test Antagonize reserpine-induced ptosis and hypothermia
Proanthocyanidin	25–50 mg/kg	Oral administration in mice	Reduce immobility period in the forced swimming test and tail suspension test Enhance 5-HT levels in hypothalamus, hypothalamus, and frontal cortex
Quercetin	20–40 mg/kg	Oral administration in mice	Prevent hyperactivation of the HPA axis
Resveratrol	20–80 mg/kg	Oral administration in mice	Decrease immobility period in the despair tests without influence on locomotor activity Enhance 5-HT and noradrenaline concentrations in the brain
	40–80 mg/kg	Oral administration in rats	Reverse less weight gain, reduce sucrose preference and deficits in the shuttle box Raise 5-HT, dopamine, and noradrenaline concentrations in brain Reduce MAO activity

Author's note: The drugs mentioned in this work are only to be taken under medical supervision; this should not be considered as a prescription or treatment guide.

Table 5.4: displays medicinal herbs that target depression (Gihyun Lee and Hyunsu Bae, 2017: This is an open-access article distributed under the Creative Commons Attribution License, which permits unrestricted use, distribution, and reproduction in any medium, provided the original work is properly cited)

Herbs	Dose	Study design	Effects and mechanisms
Camellia sinensis	5, 10, and 20 mg/kg	Oral administration in mice	Decrease immobility in the tail suspension test and forced swimming test Modulate the HPA axis
Crocus sativus	30 mg/day	Randomized controlled clinical trials	Improve the Hamilton Rating Scale for Depression
	100 mg/day	Randomized controlled clinical trials	Improve Beck Depression Inventory and Beck Anxiety Inventory Scores with rare side effects
Echium amoenum	375 mg/day	Randomized controlled clinical trials	Improve the Hamilton Rating Scale for Depression at week four of a study, however this result was not maintained at week six
Hypericum perforatum	3 × 300 mg/day	Long-term follow-up study involving 426 patients	Prevent relapse after recovery from acute depression
Piper methysticum	16 g containing 250 mg of kavalactones/day	Randomized controlled trials	Improve the Montgomery–Asberg Depression Rating Scale with no serious adverse effects and no clinical hepatotoxicity
Rhodiola rosea	1.5–6 g/kg	Oral administration in rats	Increase hippocampus 5-HT level-induced proliferation of neural stem cell, repairing the damaged neuronal cells in hippocampus
	10–20 mg/kg	Oral administration in rats	Revert decreased sucrose intake Reduce moving behavior Minimize weight gain and dysregulation of their estrous cycle
	340–680 mg/day	Randomized controlled phase III clinical trial	Improve overall depression, together with insomnia, emotional instability, and somatization, but not self-esteem with no serious side effects

Lavandula angustifolia	Lavender aromatherapy	Inhalation in rats	Inhibit depression-like behaviors in forced swimming and elevated plus-maze tests Reverse spatial memory deficits
	Lavender cream	Randomized controlled clinical trials	Reduce stress, anxiety, and depression in pregnant women
	Lavender aromatherapy	Randomized controlled clinical trials	Reduce depressive symptoms Improve the Edinburgh Postnatal Depression Scale
Nelumbo nucifera	4 g/kg	Oral administration in rats	Reverse the decreased sucrose intake and the decreased 5-H receptor binding in brain
	2.1 g/kg	Oral administration in rats	Increase struggling time and first latency time in the forced swimming test
	0–2000 mg/kg for rats 0–4000 mg/kg for beagle dogs	Oral administration in rats or beagle dogs	Improve the start latency, rearing number, grooming time, and decreased visiting counts caused by chronic mild stress in the forced swimming test No toxicity during 28 days of administration in dogs and 13 weeks of administration in rats

Author's note: The drugs mentioned in this work are only to be taken under medical supervision; this should not be considered as a prescription or treatment guide.

While the gifts of the plant kingdom show promise to target depression, Tsai et al (2012) reported in a structured literature review that the use of herbs and dietary supplements (HDS) was associated with contraindications ranging from gastrointestinal (16.4%) to neurological (14.5%) to and renal/genitourinary (12.5%) manifestations. This calls for another word of caution here in terms of using any medication to target depression.

5.5 APPS TO THE RESCUE

Given the prevalence of smartphones in our lives, Carras and team (2014) reported that 85.7% of 189 inner-city community psychiatry program patients had a mobile phone with 42% of them accessing the Internet several times a day. This brings in the use of "apps" to manage the symptoms and seek help in the case of depression.

Researchers Pung and team (2018) employed phone interviews to look at the use of mobile apps and management strategies. 16 (58% female) participants were recruited from the Target-D trial (ACTRN12616000537459: aimed at targeting depression) of whom 3/4th were depressed as assessed by the Patient Health Questionnaire (PHQ-2). Eleven of the sixteen participants used at least one of fifteen apps for the purposes of skill acquisition, social connectedness, inquisitive trial, and safety netting.

Table 5.5: Below displays few apps used by patients and their referral sources(Alison Pung, Susan Louise Fletcher, Jane Maree Gunn. Originally published in the Journal of Medical Internet Research (http://www.jmir.org), 27.09.2018: This is an open-access article distributed under the terms of the Creative Commons Attribution License (https://creativecommons.org/licenses/by/4.0/), which permits unrestricted use, distribution, and reproduction in any medium, provided the original work, first published in the Journal of Medical Internet Research, is properly cited. The complete bibliographic information, a link to the original publication on http://www.jmir.org/, as well as this copyright and license information must be included)

Type of app and app name	Referral source	Used by
Mindfulness or relaxation		
Mindfulness[a] (3 apps)	Midwife, Psychologist	P1, P5 (2 apps)
Smiling Minds	General practitioner	P4
Relax+	Psychiatrist	P14
Meditation[a]	Self-directed internet search	P16
Fitness		
Exercise[a]	Inbuilt on phone	P2
Gaming		
Ingress	Social network or friends	P3
Cognitive		

	SAM App[b]	Self-directed internet search	P7
	NeuroNation	Self-directed internet search	P7
	MindTools	Self-directed internet search	P7
	Psych Me Up	University studies	P8
	Mood Switch	University studies	P8
Social media			
	Facebook	Social network or friends	P9
Workplace well-being			
	Equipt	Workplace or organization	P11
No app used			
	N/A[c]	N/A	P6, P10, P12, P13, P15

[a]Purpose of app described where app name was not known or not specified.

[b]SAM: Self-help for Anxiety Management App.
[c]N/A: not applicable.

Five examples of what the participants shared are presented below:... you still sort of keep in contact with what sort of going on in the world. It does make you sort of happy that you still communicate and you're still out there communicating with people.

- I found that I wasn't as upset...or depressed, when I was out there playing it [Ingress] and the friends I made through it helped.
- I just like that it's there.
- I use an app on my phone which is—it's like a meditation app—it was what my psychiatrist recommended. So, yeah, I've been very disciplined when he suggested to give it a go and—yeah—and use and give it a go.
- I think ultimately though, people who need to know they're cared for by another person, a real person. An app can help but I reckon fundamentally as humans, we want someone to know we're hurting or struggling, and we need more than an app...but I think it's a good idea to have both.

The researchers state "The rise of mobile technology heralds a new era in community mental health treatment. There is growing interest from clinicians, consumers, and developers in harnessing the potential for mobile apps to address some of the barriers to traditional mental health care".

A search across app markets revealed 293 apps that offered to handle depression and mental health. Published research supporting this claim was seen in 3.41% of these apps. Affiliation to a university or institute or government body was seen in 20.48% of the apps. 74.06% of apps for depression and/or anxiety entailed no cost for downloading.

Table 5.6: Below presents apps and supporting research that help handle depression (Marshall, Dunstan and Bartik, 2019: This is an open-access article distributed under the terms of the Creative Commons Attribution License (CC BY). The use, distribution or reproduction in other forums is permitted, provided the original author(s) and the copyright owner(s) are credited and that the original publication in this journal is cited, in accordance with accepted academic practice. No use, distribution or reproduction is permitted which does not comply with these terms)

Article	App name	Design	Participants	Outcome	Time/Length	Statistical analysis
Lee and Jung (24) *	Destressify	RCT	77 participants in intervention group, 86 in waitlist control group; mean age 20.6 years	Stress, anxiety, and depression	Intervention period 4 weeks; no long-term follow-up	ANCOVA/MANCOVA; detailed in article
Christoforou et al. (25)	Agoraphobia Free	RCT	73 participants in intervention group, 69 in other app "control" group (no waitlist group), mean age 39.7 years	Anxiety (agoraphobia)	Intervention period 12 weeks; no long-term follow-up	Linear mixed model; detailed in article
Kinderman et al. (26)	Catch It	Feasibility/ pilot study	285 participants (no control group); mean age 48.0 years	Positive/negative mood	Intervention period 6 weeks maximum, but varied amongst participants; no long-term follow-up	ANOVA; detailed in article
Carey et al. (27)	Mindsurf	Feasibility/ pilot study	23 participants (no control group); no mean age given	Anxiety, depression	Intervention period 2 weeks; no long-term follow-up	ANOVA; inadequate detail in article
Kuhn et al. (28)	PTSD Coach	RCT	62 participants in intervention group, 58 in waitlist control group; mean age 39.0 years	Anxiety (posttraumatic stress disorder), depression	Intervention period 3 months; 3 months long-term follow-up	Repeated measures ANOVAs; detailed in article
Bakker et al. (29)	MoodMission	RCT	56 participants in intervention group 1, 56 in intervention group 2, 50 in intervention group 3, 64 in waitlist control group; mean age 34.0 years	Anxiety, depression	Intervention period 30 days; no long-term follow-up	ANOVA; detailed in article
Roepke et al. (30)	SuperBetter	RCT	93 participants in intervention group 1, 97 in intervention group 2, 93 in waitlist control group; mean age 40.2 years	Anxiety, depression, life satisfaction	Intervention period 4 weeks; 6 weeks long-term follow-up	Hierarchical Linear Modeling; detailed in article
Stiles-Shields et al. (31)	Thought Challenger	RCT	10 participants in intervention group 1, 10 in intervention group 2, 10 in waitlist control group; no mean age given	Depression	Intervention period 6 weeks; 4 weeks long-term follow-up	Repeated measures ANOVA; detailed in article
Flett et al. (32) *	Smiling Mind	RCT	58 participants in intervention group, 67 in placebo control group; mean age 20.1 years	Stress, anxiety, depression, flourishing	Intervention period 40 days; no long-term follow-up	Multiple regression; detailed in article
Flett et al. (32) *	Headspace	RCT	67 participants in intervention group, 67 in placebo control group (same control group as above study); mean age 20.1 years	Stress, anxiety, depression, flourishing	Intervention period 40 days; no long-term follow-up	Multiple regression; detailed in article

**Independent research. RCT, Randomized controlled trial; ANOVA, Analysis of variance; ANCOVA, Analysis of covariance; MANCOVA, Multivariate analysis of co-variance.*

Overall, apps have made a foray into the realm of depression. However, an expansion of research is warranted with increased participation from health-care givers and governments.

To summarize, given that depression is a complex condition caused by a combination of genetic and environmental factors, its treatment too warrants careful screening, diagnosis, and treatment according to the condition.

5.6 REFERENCES

1. American Psychiatric Association: Practice guideline for the treatment of patients with major depressive disorder, ed 3. Am J Psychiatry 2010;167(suppl):1–118.
2. Giovanni A. Fava. Rational Use of Antidepressant Drugs. PsychotherPsychosom2014;83:197–204 DOI: 10.1159/000362803.
3. Andrea Cipriani, Toshi A Furukawa*, Georgia Salanti*, Anna Chaimani, Lauren Z Atkinson, Yusuke Ogawa, Stefan Leucht, Henricus G Ruhe, Erick H Turner, Julian P T Higgins, Matthias Egger, Nozomi Takeshima, Yu Hayasaka, Hissei Imai, Kiyomi Shinohara, Aran Tajika, John P A Ioannidis, John R Geddes. Comparative efficacy and acceptability of 21 antidepressant drugs for the acute treatment of adults with major depressive disorder: a systematic review and network meta-analysis. Lancet 2018; 391: 1357–66
4. Gabriel FC, de Melo DO, Fráguas R, Leite-Santos NC, Mantovani da Silva RA, Ribeiro E (2020) Pharmacological treatment of depression: A systematic review comparing clinical practice guideline recommendations. PLoS ONE 15(4): e0231700. https://doi.org/10.1371/journal.pone.0231700
5. Heddaeus D, Steinmann M, Daubmann A, Härter M, Watzke B (2018) Treatment selection and treatment initialization in guideline-based stepped and collaborative care for depression. PLoS ONE 13(12): e0208882. https://doi.org/10.1371/journal.pone.0208882
6. Coupland, C., Hill, T., Morriss, R. et al. Antidepressant use and risk of adverse outcomes in people aged 20–64 years: cohort study using a primary care database. BMC Med **16,** 36 (2018). https://doi.org/10.1186/s12916-018-1022-x
7. Moore TJ, Mattison DR. Adult utilization of psychiatric drugs and differences by sex, age, and race. JAMA Intern Med. 2017;177:274–5.
8. Gihyun Lee, Hyunsu Bae, "Therapeutic Effects of Phytochemicals and Medicinal Herbs on Depression", BioMed Research International, vol. 2017, Article ID 6596241, 11 pages, 2017. https://doi.org/10.1155/2017/6596241.
9. Moshiri E, Basti AA, Noorbala AA, Jamshidi AH, Hesameddin Abbasi S, Akhondzadeh S. Crocus sativus L. (petal) in the treatment of mild-to-moderate depression: a double-blind, randomized and placebo-controlled trial. Phytomedicine. 2006 Nov;13(9-10):607-11. doi:

10.1016/j.phymed.2006.08.006. Epub 2006 Sep 18. PMID: 16979327.

10. KaijunNiu, Atsushi Hozawa, Shinichi Kuriyama, Satoru Ebihara, Hui Guo, Naoki Nakaya, Kaori Ohmori-Matsuda, Hideko Takahashi, Yayoi Masamune, Masanori Asada, Satoshi Sasaki, Hiroyuki Arai, Shuichi Awata, Ryoichi Nagatomi, Ichiro Tsuji, Green tea consumption is associated with depressive symptoms in the elderly, The American Journal of Clinical Nutrition, Volume 90, Issue 6, December 2009, Pages 1615–1622, https://doi.org/10.3945/ajcn.2009.28216
11. Conrad P, Adams C. The effects of clinical aromatherapy for anxiety and depression in the high risk postpartum woman - a pilot study. Complement Ther Clin Pract. 2012 Aug;18(3):164-8. doi: 10.1016/j.ctcp.2012.05.002. Epub 2012 Jun 27. PMID: 22789792.
12. Tsai HH, Lin HW, Simon Pickard A, Tsai HY, Mahady GB. Evaluation of documented drug interactions and contraindications associated with herbs and dietary supplements: a systematic literature review. Int J Clin Pract. 2012 Nov;66(11):1056-78. doi: 10.1111/j.1742-1241.2012.03008.x. PMID: 23067030.
13. Colder Carras, M., Mojtabai, R., Furr-Holden, C. D., Eaton, W., & Cullen, B. A. (2014). Use of mobile phones, computers and internet among clients of an inner-city community psychiatric clinic. Journal of psychiatric practice, 20(2), 94–103. https://doi.org/10.1097/01.pra.0000445244.08307.84
14. Pung, A., Fletcher, S. L., & Gunn, J. M. (2018). Mobile App Use by Primary Care Patients to Manage Their Depressive Symptoms: Qualitative Study. Journal of medical Internet research, 20(9), e10035. https://doi.org/10.2196/10035
15. Marshall, J. M., Dunstan, D. A., &Bartik, W. (2019). The Digital Psychiatrist: In Search of Evidence-Based Apps for Anxiety and Depression. Frontiers in psychiatry, 10, 831. https://doi.org/10.3389/fpsyt.2019.00831

CHAPTER 6

YOGA AND PRANAYAMA FOR DEPRESSION

CONTENTS

6.1 YOGA: WHAT DOES RESEARCH SAY?

The term "yoga" refers to the unity of mind and body practiced in India that has now received the limelight by many parts of the world. Dating back to 3000 years ago, the roots of this practice can be attributed to Patanjali who developed 8 limbs or aspects of this practice. The positions or asanas are representative of only one aspect! Scientific studies are uncovering various benefits of this approach on osteoarthritis, hypertension, lymphoma, asthma, multiple sclerosis, irritable bowel syndrome, and depression.

(Schreiber and Niemeier, 2013; Shohani et al, 2018)

According to Groessl et al (2015), "*Clearly, in cases where yoga does not offer significant relief from physical disease, it can still offer some measure of relief from suffering. In this sense, yoga can provide a different way of looking at pain and suffering, which in itself can potentially alleviate some suffering. The basic realization that "I" am not my body, or my thoughts, or my sensations of pain, is itself healing or liberative from a certain kind of existential suffering*".

Let us see what research has to say about the use of yoga.

Early work in 2004 explored the changes in the moods of young adults who were mildly depressed due to a short-term Iyengar yoga course. Two Iyengar yoga classes each week for 5 consecutive weeks (1-hour duration) were administered to 18-29-year-old volunteers (n=28) with no previous exposure to yoga or on psychiatric diagnoses or treatments. By the middle of the course, the self-reported depressive symptoms reported an improvement that was maintained till the end. The levels of morning cortisol were elevated in the participants of the yoga course. In the words of the scientists, "*These findings provide suggestive evidence of the utility of yoga asanas in improving mood and support the need for future studies with larger samples and more complex study designs to more fully evaluate the effects of yoga on mood disturbances*".

A study published in 2011 examined the effects of yoga in 81 undergraduate students (their age was 18 and above) with mild to moderate depression, anxiety, or stress. Among the participants who were divided into integrated yoga, yoga as exercise, and controls, those who participated in the integrated and exercise yoga groups displayed lowered depression with increased hopeful emotions albeit over time. The levels of salivary cortisol and anxiety showed a dip from the start to the end of the study opening up the avenue for the promise of yoga to treat depression. According to the authors, "*Yoga, practiced in a more integrated form, ie, with an ethical and spiritual component, may provide additional benefits over yoga practiced as an exercise regimen*" (Smith et al, 2011).

6.2 YOGA: A RESEARCH SURVEY

A National Center for Health Statistics report published in 2015 presented the national estimates of the use of yoga along with other aspects such as natural supplementsas part of complementary health approaches to maintain a "wellness lifestyle" among U.S. adults in 2012. The basis of the study was the 2012 Adult Alternative Medicine supplement to NHIS (National Health Interview Survey) that was entailed the answers of 34,525 adults to a questionnaire followed by statistical analyses. The use of yoga was found to be for wellness purposes by a factor of five over that of treating a health issue.

Table. Questionnaire items to capture wellness-related reasons for use and perceived health outcomes from use of complementary health approaches: National Health Interview Survey, 2012

Reasons for and outcomes from using complementary health approaches	Survey question
General wellness or disease prevention	Did you [see a practitioner for/use modality] for any of these reasons? For general wellness or general disease prevention?
Improve immune function	Did you [see a practitioner for/use modality] for any of these reasons? To improve your immune function?
Improve energy	Did you [see a practitioner for/use modality] for any of these reasons? To improve your energy?
Focuses on the whole person—mind, body and spirit	During the past 12 months, did you [see a practitioner for/use modality] for any of these reasons? It focuses on the whole person—mind, body, and spirit?
Improve memory or concentration	Did you [see a practitioner for/use modality] for any of these reasons? To improve your memory or concentration?
Exercise more regularly	Did [seeing a practitioner for/using modality] motivate you to exercise more regularly?
Eat healthier	Did [seeing a practitioner for/using modality] motivate you to eat healthier?
Cut back or stop drinking alcohol	Did [seeing a practitioner for/using modality] motivate you to cut back or stop drinking alcohol?
Cut back or stop smoking cigarettes	Did [seeing a practitioner for/using modality] motivate you to cut back or stop smoking cigarettes?
Reduced stress or relaxation	Did [seeing a practitioner for/using modality] lead to any of these outcomes? Help to reduce your stress level or to relax?
Improved sleep	Did [seeing a practitioner for/using modality] lead to any of these outcomes? Help you to sleep better?
Felt better emotionally	Did [seeing a practitioner for/using modality] lead to any of these outcomes? Make you feel better emotionally?
Eased coping with health problems	Did [seeing a practitioner for/using modality] lead to any of these outcomes? Make it easier to cope with health problems?
Improved overall health and made you feel better	Did [seeing a practitioner for/using modality] lead to any of these outcomes? Improve your overall health and make you feel better?

NOTE: Response options for questionnaire items are yes or no.

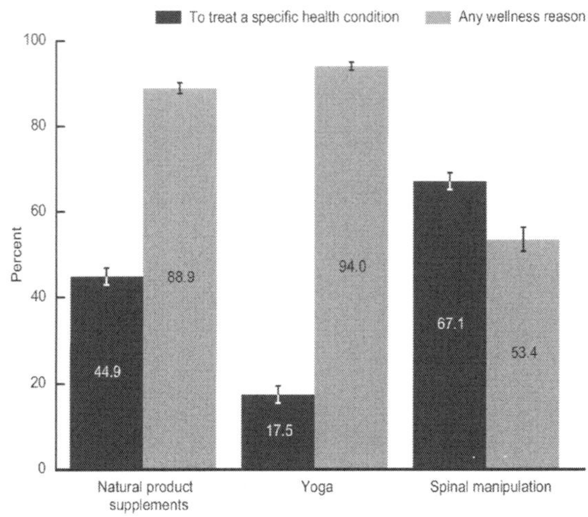

Figure 6.1: Above shows ***TOP***: The format of the questionnaire; ***BOTTOM***: The reason for using yoga by the participants(Stussman et al, 2015; *All material appearing in this report is in the* public domain *and may be reproduced or copied without permission; citation as to source, however, is appreciated*).

More than 2/3rdof the participants reported that yoga augmented the energy or it helped to focus the mind, spirit, and body. 30% of yoga practitioners expected an improvement in memory and concentration.

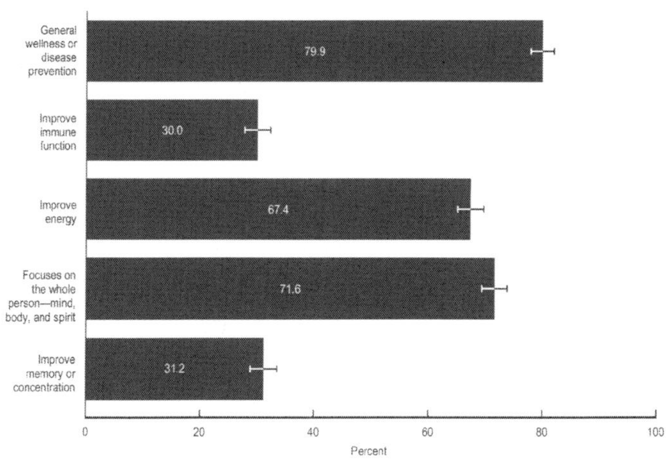

Figure 6.2: Above displays these findings (Stussman et al, 2015; *All material appearing in this report is in the public domain and may be reproduced or copied without permission; citation as to source, however, is appreciated*).

Estimating the self-reported perceived health behavior revealed that the use of yoga helped to cut down on unhealthy patterns: drinking or smoking more than the effect of natural supplements or spinal manipulation!

The results were:

- For motivation to be more regular to exercise: the positive effect of yoga was found to be 63.2% against 16.0% for natural supplements and 21.7% for spinal manipulation.
- For eating healthy: the rate of yoga was 42.6% against 21.9% for natural supplements and 10.7% for spinal manipulation.
- To limit alcohol consumption, the rate of yoga was 12.4% against 4.6% for natural supplements and 2.7% for spinal manipulation.
- To limit smoking, the rate of yoga was 25.2% against 5.6% for natural supplements and 6.5% for spinal manipulation.

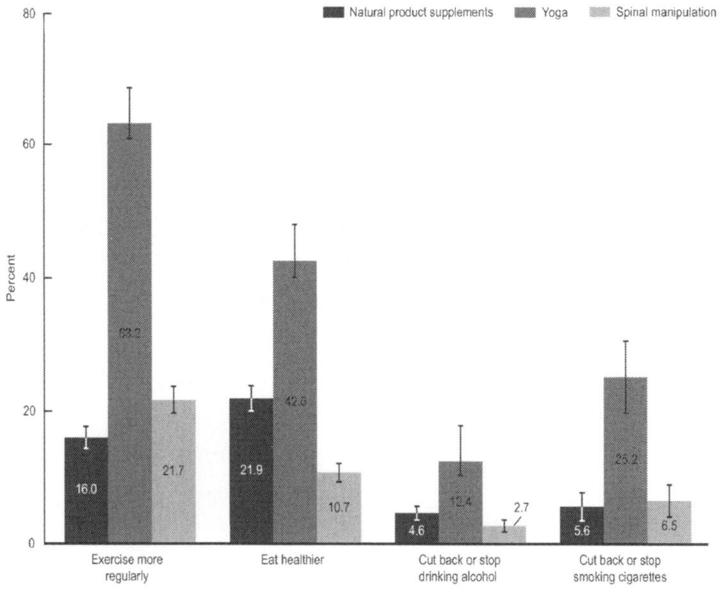

Figure 6.3: Above displays these results (Stussman et al, 2015; *All material appearing in this report is in the public domain and may be reproduced or copied without permission; citation as to source, however, is appreciated*).

Looking at the results, the use of yoga for 12 months resulted in more than 80% of the users feeling less stressed. 67.2% of those who practiced yoga felt emotionally better over 21.4% of users of natural supplements and 27.3% of those who used spinal manipulation.

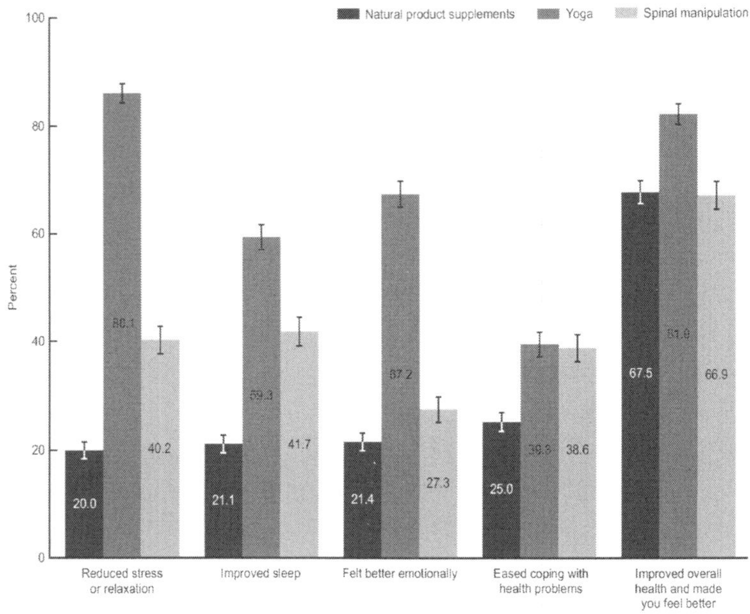

Figure 6.4: The outcomes of following yoga for 12 months (Stussman et al, 2015; *All material appearing in this report is in the public domain and may be reproduced or copied without permission; citation as to source, however, is appreciated*).

Described as "*the most comprehensive, nationally representative data available describing wellness-related reasons and self-reported perceived health outcomes for adults using complementary health approaches*", this 2015-National Health Statistics Report outlined that close to 2/3rd of the yoga practitioners felt more motivated to exercise more regularly while feeling emotionally better and less stressed (Stussman et al, 2015).

This becomes significant in a world where one is constantly on the quest to feel motivated and less stressed. With this backdrop, this chapter discusses research articles that assessed the effect of yoga and meditation on depression.

6.3 YOGA FOR DEPRESSION IN CANCER PATIENTS

Research has shown that cancer patients often experience depression and anxiety, for example, 50% of Indian cancer patients in Bangalore undergoing radiation treatment experienced depression (Chaturvedi et al, 1996). Such depression adversely impacts the quality of life and influences the treatment-related distress in these patients. In order to address these aspects of health, yoga has been assessed in several studies.

Chandwani et al (2014) assessed the effects of a yoga program involving: "1) preparatory warm-up synchronized with breathing; (2) selected postures, or asana (forward-, backward-, and side-bending asanas in sitting and standing position, cobra posture, crocodile, and half-shoulder-stand with support); (3) deep relaxation (supine posture); (4) alternate-nostril breathing, or pranayama; and (5) meditation" as taught by the Vivekananda Yoga Anusandhana Samsthana (VYASA) on breast cancer patients undergoing treatment. Random assignment of patients (stages 0 to III) breast cancer was done with 53 patients receiving yoga for 6 weeks; 56 received stretching exercises, and 54 were controls. The analyses of the quality of life (QOL) factors revealed an improvement due to the yoga program with a steep cortisol profile suggestive of lowered stress.

Of 98 stage II and III breast cancer patients: 45 received an integrated yoga program: " asanas, breathing exercises, Pranayama (voluntarily regulated nostril breathing), meditation, and yogic relaxation techniques with imagery" that also included Vedic chanting while 53 patients received supportive counseling. Both treatments were over 24 weeks and also involved conventional chemo/radiotherapy. The depression scores, as assessed by the Beck's Depression Inventory (BDI) were lowered by yoga by 28.1% during radiotherapy, 28.5% after radiotherapy, 42% post-surgery, 39.5% during chemotherapy, and 29.2% after chemotherapy as against controls. The researchers stated that "*Overall, the study shows benefit finding for reducing self-reported depression scores in operable breast cancer patients undergoing cancer directed treatment*" (Rao et al, 2015).

Given that about 40% of adult cancer patients resort to complementary and alternative medicine (Horneber et al, 2012), a feasibility study of augmenting the health-related quality of life (HRQoL) by yoga was published by Marcy McCall and team in 2015. 15 participants (11 females and 4 males) undergoing therapy for breast, prostate, colorectal, brain, and blood and lung cancers were enrolled from the Vancouver Centre of the British Columbia Cancer Agency (BCCA) of which the study was completed by 12: (on account of a scheduling conflict and surgery).

Table 6.1: The details of the patients (Marcy McCall et al, 2015; This is an open-access article distributed under the Creative Commons Attribution License, which permits unrestricted use, distribution, and reproduction in any medium, provided the original work is properly cited): n.r: not reported

Participant	Age at consent	Gender		Tumour location	Stage of cancer					Conventional treatment status				
		Male	Female		I	II	III	IV	n.r	Chemotherapy	Radiotherapy	Hormone therapy	Active surveillance	After surgery
A1	66	1		Tongue	1					1	1			
A2	60		1	Breast		1					1	1		
A3	51		1	Breast		1				1	1	1		
A4	40		1	Breast			1			1				
A5	41		1	Colorectal-lung	1								1	
Group A total		1	4		2	2	1	0	0	3	3	2		0
Average	51.6													
B1	72	1		Prostate			1				1		1	

B2	61		1	Abdomen				1				1	1	1
B3	43		1	Breast	1					1		1		
B4	51		1	Colorectal			1			1	1			
B5	59		1	Skin (fore-arm)					1		1			
roup B total		1	4		1	0	2	1	1	2	3	2	2	1
Aver-age	57.2													
C1	57	1		Prostate	1								1	
C2	52		1	Breast			1			1	1		1	
C3	33		1	Brain			1			1	1			
C4	45		1	Breast		1						1		1
C5	38	1		Blood					1					
Group C total		2	3		1	1	2	0	1	2	2	1	2	1
Aver-age	45													

Overall	**4**	**11**	**4**	**3**	**5**	**1**	**2**	**7**	**8**	**5**	**5**	**2**		
Aver-age	**51.3**													

Table 6.2: There were 3 yoga intervention programs assigned to the patients for 28 days: the scheme and approach utilized below(Marcy McCall et al, 2015; *This is an open-access article distributed under the Creative Commons Attribution License, which permits unrestricted use, distribution, and reproduction in any medium, provided the original work is properly cited*)

	a) QOL-CA questionnaire completed by consenting participants (online or paper)
	b) Semi-structured, self-reported survey (Y-ACT) to elicit information regarding at-home yoga adherence, patient experiences, utility/dysutility of yoga, patient attitude and barriers to practising yoga (online or paper)
	1) 30-minute training session of pranayama (breathing techniques) administered by a certified yoga instructor in a group setting, given own yoga mat and book to encourage home-based practice, social gathering and tea offered to encourage group discussion
	2) 15-minute orientation and education session delivered by a certified yoga instructor, includes identification and list of community-based and online options to practise yoga

	3) 30-minute of asana training with 15-minute orientation of free and unlimited access to online yoga classes (http://www.myyogaonline.com)
	4) 120-minute yoga workshop administered by a certified yoga instructor in a group setting, orientation and education session to include pranayama, dhyana, asana (physical postures), social gathering and tea offered to encourage group discussion
	5) 3 × 60-minute yoga classes (per week) including pranayama, asana and dhyana administered by a certified yoga instructor in a group setting, social gathering and tea offered to encourage group discussion

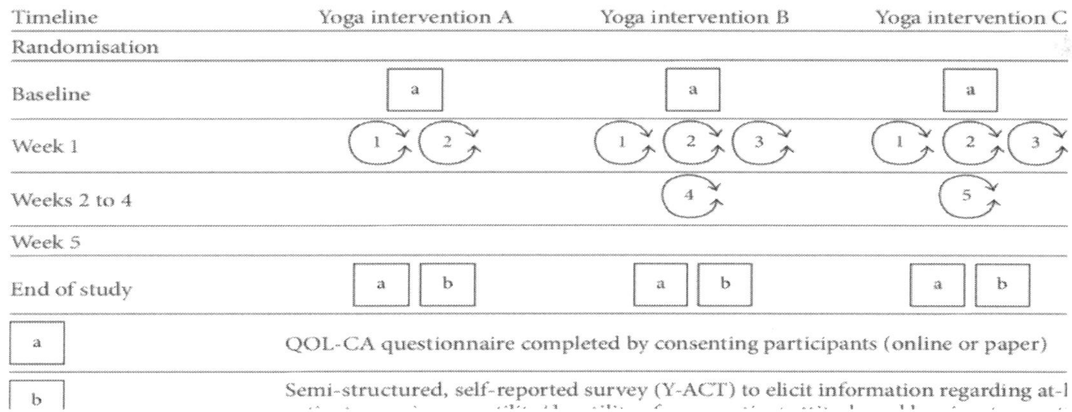

Session number	Section title	Duration	Key components	Lesson plan outline (i.e., how many minutes of what component?)
Group A: Low-dose yoga intervention				
1	Introduction to yoga	15 min	What is yoga? How and why practice yoga? Discuss resources and how to access community-based or affordable yoga	What is Yoga? Connection of mind and body, with others and your environment; awareness. Why? To create ease, balance, and awareness; calm the mind; tool to manage symptoms and feelings; stress management. No one way to practice yoga; many styles and variations. The right way is your way. Listen to your body, always, over and above any teacher. When taking class feel free to always rest when you need it or modify postures. You are the only one who knows how your body feels. If it does not feel right, it is not right. You can always talk to the teacher before class and let them know how you are feeling. Give them as little/much info as you want. This will help them to help you. Resource list to access other yoga

1	Yoga training	30 min	Pranayama techniques; Satsang	4 minutes: introduction and overview of class direction, check-in with students (either individually or collectively) 4 minutes: breath witness/awareness 12 minutes: pranayama 10 minutes: Satsang, tea, and group discussion Note: pranayama to be done in Savasana or supported restorative, to be picked by teacher. Teacher can pick pranayama based on the feel of the room and energy level of students. Please pick from the following list. To use belly breath, sectional breathing, sighing out breathing, brahmari, and sitkari Not recommended: breath holding (causes dizziness) and nadi shodhana
Group B: *Medium-dose yoga intervention*				
1	Introduction to yoga and online membership	15 min	What is yoga? How and why practice yoga? Introduce membership to http://www.myyogaonline.com and give pamphlet for community-based resources	Similar to Group A, with addition of online yoga orientation Talk about "myyogaonline" website; instruct to search on the left side of screen for Hatha or restorative yoga classes or specific classes for anxiety, breathing, detoxification, gentle, relaxation, or therapeutics. Emphasise that the student only does what they feel comfortable with. One of the offerings of yoga is learning to listen to your own body and then having the respect to listen to it. If it feels good, great. If it feels bad, adjust/modify or stop.
1	Yoga training	30 min	Pranayama techniques	Same as low-dose (see above)

1	Yoga training	30 min	Asana practice and safety training	All very gentle asana 10 minutes: restorative yoga asana, floor-based 15 minutes: active yoga asana, upright and on feet (muscle retention and very easy balancing) 10 minutes: restorative yoga asana, floor-based 10 minutes: guided meditation, dhyana
2	Yoga workshop	120 min	Pranayama, asana, and dhyana techniques	10 minutes: welcome, check-in 10 minutes: witness/awareness and pranayama 20 minutes: restorative yoga asana, floor-based 15 minutes: active yoga asana, upright and on feet 20 minutes: restorative yoga asana, floor-based 10 minutes: guided meditation, dhyana 10 minutes: Satsang, tea, and group discussion 25 minutes to be used at teachers discretion based on class
Group C: High-dose yoga intervention				
1	Introduction to yoga and online membership	15 min	What is yoga? How and why practice yoga? Introduce membership to http://www.myyogaonline.com and give pamphlet for community-based resources	Same as medium-dose (see above)
1	Yoga training	30 min	Pranayama techniques; Satsang	Same as low- and medium-dose (see above)
1	Yoga training	30 min	Asana practice and safety training	Same as medium-dose (see above)

2 to 12	Yoga training	60 min	Pranayama, asana, and dhyana techniques; Satsang	3 minutes: welcome, check-in 2 minutes: seated pranayama exercise 10 minutes: restorative yoga asana, floor-based 15 minutes: active yoga asana, upright, and on feet 10 minutes: restorative yoga asana, floor-based 10 minutes: guided meditation, dhyana 10 minutes: Satsang, tea, and group discussion

Table 6.3: The expenditure involved was $255 for Group A, $325 for Group B and $500 for Group C. Following data analyses, a positive experience was reported by all the 12 participants. The participant feedback (***TOP***) and the analyses of the QOL-CA/CS survey (to assess the quality of life of cancer patients-***BOTTOM***) (—: not reported; SD = standard deviation; CI = confidence interval; complete data sets only (); subgroups (); Marcy McCall et al et al, 2015; *This is an open-access article distributed under the Creative Commons Attribution License, which permits unrestricted use, distribution, and reproduction in any medium, provided the original work is properly cited*)

Participant	**Number of total yoga sessions**	**Number of online yoga sessions practised**	**Amount of satisfaction participating in yoga (0 = none, 10 = very)**	**Amount of satisfaction doing yoga with teacher (0 = none, 10 = very)**	**Amount of satisfaction using online yoga**	**Felt not being offered enough yoga**	**Would prefer more online yoga**	**Right amount of yoga**	**Too much yoga**	**Utility of yoga (0 = none, 10 = very)**	**Disutility of yoga (0 = none, 10 = very)**
A1	1	0	8	7	—					8	5
A3	6	4	8	8	8	1				8	0
A4	1	0	1	1	5	1				8	0
A5	8	6	10	10	9	1				10	2
Average	**4**	**2.5**	**6.8**	**6.5**	7.3					**8.5**	**1.8**

Group total						**3**	**0**	**0**	**0**		
B1	4	0	8	9	0	1				8	0
B2	2	0	7	9	5	1				7	2
B3	10	0	5	10	0	1				8	4
B5	4	0	8	8	—	1				6	1
Average	**5**	**0**	**7**	**9**	**1.7**					**7.3**	**1.8**
Group total						**4**	**0**	**0**	**0**		
C1	10	0	9	9	—				1	8	0
C2	24	0	10	10	—	1				10	7
C3	5	0	6	7	7			1		8	2
C4	12	0	9	9	0	1				10	
Average	**12.8**	**0**	**8.5**	**8.8**	**3.5**					**9**	**2.3**
Group total						**2**		**1**	**1**		

Summary of QOL-CA/CS subscale results.

Physical Wellbeing Subscale				Psychological Wellbeing Subscale			
Participants	Baseline	Post-inter-vention	Mean Dif-ference	Partic pants	Baseline	Post-inter-vention	Mean Dif-ference
Group A				Group A			
Average (SD)	5.82 (1.27)	5.19 (3.05)	−0.63 (2.66)	Average (SD)	4.85 (1.71)	5.73 (2.18)	0.88 (1.09)
95% CI	3.80 to 7.84	0.34 to 9.77	−3.60 to 4.86	95% CI	2.13 to 7.57	2.24 t 9.19	−2.61 to 0.85
Group B				Group B			
Average (SD)	7.21 (2.25)	7.88 (1.76)	0.67 (0.78)	Average (SD)	5.97 (1.20)	5.51 (1.08)	−0.47 (0.96)
95% CI	3.64 to 10.78	5.08 to 10.68	−0.57 to 1.91	95% CI	4.07 to 7.87	3.79 to 7.23	−2.00 to 1.06
Group C				Group C			
Average (SD)	7.32 (2.59)	6.60 (2.81)	−0.72 (1.54)	Average (SD)	4.44 (1.74)	4.89 (1.29)	0.45 (0.45)
95% CI	3.20 to 11.44	2.13 to 11.07	−3.18 to 1.74	95% CI	1.67 to 7.21	−1.56 to 2.54	−0.27 to 1.17
Overall				Overall			
Avera e (SD)	6.78 (2.04)	6.55 (2.59)	−0.23 (1.76)	Average (SD)	5.09 (1.57)	5.37 (1.48)	0.29 (0.99)
95% CI	5.49 to 8.07	4.90 to 8.20	−1.35 to 0.89	95% CI	4.09 to 6.09	4.43 to 6.31	−0.34 to 0.92

Social Wellbeing Subscale				Spiritual Wellbeing Subscale			
Partici ants	Baseline	Post-inter-vention	Mean Dif-ference	Participants	Baseline	Post-inter-vention	Mean Dif-ference
Group A				Group A			
Average (SD)	4.94 (1.34)	5.29 (2.74)	0.35 (1.63)	Average (SD)	3.82 (1.18)	5.78 (1.97)	1.96 (1.15)
95% CI	2.80 to 7.08	0.93 to 9.65	−2.24 to 2.94	95% CI	1.94 to 5.70	2.65 to 8.91	0.13 to 3.79
Group B				Group B			
Average (SD)	6.29 (1.85)	6.47 (1.92)	0.19 (1.02)	Average (SD)	5.14 (1.35)	5.07 (1.56)	−.07 (1.74)
95% CI	3.35 to 9.23	3.42 to 9.52	−1.43 to 1.81	95% CI	3.00 to 7.28	2.59 to 7.55	−2.70 to 2.84
Group C				Group C			
Average (SD)	5.07 (2.06)	4.86 (1.86)	−0.21 (0.78)	Average (SD)	5.86 (1.62)	6.68 (1.78)	0.83 (0.55)
95% CI	1.79 to 8.35	1.90 to 7.82	−1.45 to 1.03	95% CI	3.28 to 8.44	3.85 to 9.51	−0.05 to 1.66
Overall				Overall			
Average (SD)	5.43 (1.73)	5.54 (2.13)	0.11 (1.11)	Average (SD)	5.18 (1.68)	5.85 (1.75)	0.67 (1.63)
95% CI	4.43 to 6.53	4.19 to 6.89	−0.59 to 0.81	95% CI	4.11 to 6.25	4.74 to 6.96	−0.42 to 1.70

In a nutshell, this feasibility study showed the suitability of yoga as an intervention for mental health in cancer treatment. The researchers state, "*Members in all groups expressed likability for yoga intervention, and perceived yoga practice as helpful for relaxation and it provided an opportunity for positive social interaction with other cancer patients*".

In an interesting study, qualitative analyses of how cancer survivors and their caregivers felt of the impact of yoga were reported. Mackenzie and team (2016) assessed the experiences of 20 cancer survivors [breast cancer, cervical cancer, colorectal cancer, lymphoma, ovarian cancer, and prostate cancer] and 5 caregivers, all attendees of the Yoga Thrive program in terms of *"(1) safety and shared understanding; (2) cancer-specific yoga instruction; (3) benefits of yoga participation; (4) mechanisms of yoga practice*".

Looking at the first aspect of understanding, the responses of the participants ranged are presented below (Mackenzie et al, 2016: *This is an open-access article distributed under the Creative Commons Attribution License, which permits unrestricted use, distribution, and reproduction in any medium, provided the original work is properly cited*):

"*I wasn't comfortable telling people that I had cancer. I don't want to go into another class and tell them, 'Well I have had cancer and I have these injuries.'* I didn't want that. I didn't feel that with (Yoga Thrive). It was like everybody is in the same boat"

"*Yoga was so restorative. It was just wonderful and such a good step to come back into yoga because I had done it previously*"

"What I think I got was you come with your own burden as a family member around the responsibility and what it looks like in the future and what's that going to mean. The benefit I didn't appreciate before. I was thinking I was helping just by signing us up. I think it relieves your own stress by normalizing some of that discussion around cancer if it came up. People who were living with it in different ways, seeing other people cope with it. I found it relieved my own stress around being the support system for someone who was living with cancer. That was an added benefit. I thought I was helping him (by co-attending the program), but it really does help as a partner in that sort of progression of living with cancer. I didn't anticipate that. That's why I enjoy it too, because it relieves some stress and takes some of that burden away to feel like I am the only option… it has provided in the same way for me as it has for him. I wasn't expecting that"

You're never discouraged – it's something I have chosen to do: something I have control over. It is at a time in your life when you have very little control over anything else that could happen physically to you. So that to me was huge. I make the decision, "I am attending. I can do this." It was getting that control back of my body I guess really... I don't know how you can get that across to people?

"I really wanted to find something that was just going to be for me – to help me feel better, to feel stronger. It had nothing to do with cancer. That's what I love about this yoga"

"I leave in a good mood every single time. I have never left yoga in a cranky mood ever."

"I think I originally enrolled for the physical benefits. The mental benefits came later." One participant stated, *"It took me, I would say, the majority of the time to just calm down and take in more of the therapeutic aspect and enjoy it. At the beginning I just kept thinking, 'OMG this isn't getting me anywhere'* and I couldn't focus. I just wanted to *get more physical, thinking that's what I needed...At the end I thought, 'Wow that was good!'* I just calmed down and my whole perspective changed. I started enjoying the class." These benefits were often also experienced outside of class, "When I am worried or thinking or going off, I will stop and just try and breathe. I try to incorporate what I have learnt in the class." Others posited emotional benefits, "If I am feeling emotional or some stress in my life before I would just keep spinning and spinning and getting tighter and tighter. Now it's like I have learned a tool to help me relax and let go of a lot of stuff. It's tremendously emotionally helpful. The techniques, tools you learn in yoga are applicable to everyday life: everywhere you go, everything you do."

"Yoga was the first time I realized I could lower my blood pressure by the way I breathed. I could take a breath and feel the stress drain away. The first time I saw that mind-body connection very clearly and I have used it actually quite often over the last three or four years.

I do have some lower back problems and I know I get tense. One thing I have learned to do is to breathe into those spots, which is how your get your muscles to relax. And I learnt that from the yoga program, nowhere else.

There was more confidence overall in myself. With confidence came an overall strength. It's the can do attitude that starts evolving from that as well. It's like a dominos thing - or a snowball really. It just keeps building and building and building...."

To summarize, the authors concluded that the practice of yoga as a community shifted the focus of cancer survivors from the disease towards the practice. Most importantly, the ability of a survivor to "take charge" once more of their body and thoughts is encouraging!

6.4 YOGA, DEPRESSION, AND THE ELDERLY

Let us see an example of how yoga was shown to augment the physical function and health-related quality of life (HRQoL) in elderly individuals. Researchers Sivaramakrishnan and team (2019) published an article (PROSPERO (Registration number: CRD42016038052) in the *International Journal of Behavioral Nutrition and Physical Activity*looking at yoga and physical functions in elderly adults given that despite the health benefits and WHO recommendation of physical activity in this population, the numbers actually doing physical functions are not up to the mark. Following querying for "yoga" and "older adults (here 60 years and above)" in Medline, PsycInfo, CINAHL Plus, Scopus, Web of Science, Cochrane Library, Embase, SPORTDiscus, AMED, ProQuest Dissertations & Theses Global databases, data extraction and analyses resulted in 27 records from 22 Randomized Controlled Trials.

Table 6.4: The details of these studies (Sivaramakrishnan et al, 2019: *Open Access; This article is distributed under the terms of the Creative Commons Attribution 4.0 International License (http://creativecommons.org/licenses/by/4.0/), which permits unrestricted use, distribution, and reproduction in any medium, provided you give appropriate credit to the original author(s) and the source, provide a link to the Creative Commons license, and indicate if changes were made. The Creative Commons Public Domain Dedication waiver (http://creativecommons.org/publicdomain/zero/1.0/) applies to the data made available in this article, unless otherwise stated*)

Study id Country	Participants (number, mean age (SD))	Intervention group (type, frequency, session duration, length of intervention)	Control group(s)	Outcome measures
Bezerra (2014) [75] Brazil	*N*=36 Mean age (SD): yoga group=63.1 (13.3); control group=61.0 (6.9) years.	Yoga, 3 times a week, 65 min per session, 12 weeks	1. IC: Control group	Physical function: Body weight

Study id Country	Participants (number, mean age (SD))	Intervention group (type, frequency, session duration, length of intervention)	Control group(s)	Outcome measures
Bonura (2014) [63] USA	*N*=98 Mean age (SD): 77.04 (7.28) years	Chair yoga, 1 session a week, 45 min per session, 24 weeks	1. AC: Chair exercise 2. IC: Wait-list control	HRQoL: Anger, anxiety, depression, well-being, general self-efficacy, self-efficacy for daily living, self-control
Bethany (2005) [55] USA	*N*=42 Mean age (SD): 83.14 (4.84) years	Chair yoga, 3 times a week, 30 min per session, 6 weeks	1. AC: Chair aerobics 2. AC: Walking programme 3. IC: Game playing	HRQoL: Stress frequency, stress severity, state anxiety, depression, motivational factors to exercise
Chen (2008) [53] Taiwan	*N*=176 Mean age (SD): Complete Silver Yoga=65.81 (4.34); Shortened Silver Yoga=68.08 (6.32); control group=72.42 (6.04) years	1. Complete Silver Yoga with meditation, 3 times a week, 70 min per session, 24 weeks 2. Shortened Silver Yoga, 3 times a week, 55 min per session, 24 weeks	1. IC: Wait-list control	Physical function: Bodyweight, BMI, body fat percentage, lower body flexibility, upper limb strength, lower limb strength, balance, walking speed, range of motion: flexion and abduction of shoulder and hip joints on both sides
Chen (2009) [61] Taiwan	*N*=128 Mean age (SD): 69.20 (6.23) years	Silver Yoga, 3 times a week, 70 min per session, 24 weeks	1. IC: Wait-list control	HRQoL: Sleep quality, depression, perceived mental and physical health

Study id Country	Participants (number, mean age (SD))	Intervention group (type, frequency, session duration, length of intervention)	Control group(s)	Outcome measures
Chen (2010)1 [60] Taiwan	N=55 Mean age (SD): 75.40 (6.70) years	Silver Yoga, 3 times a week, 70 min per session, 24 weeks	1. IC: Wait-list control	HRQoL: Sleep quality, depression, perceived mental and physical health
Chen (2010)2 [59] Taiwan	N=55 Mean age (SD): 75.40 (6.70) years	Silver Yoga, 3 times a week, 70 min per session, 24 weeks	1. IC: Wait-listcontrol	Physical function: Body weight, BMI, body fat percentage, cardiopulmonary fitness, upper body flexibility, lower body flexibility and range of motion: flexion and abduction of shoulder and hip joints on both sides, upper limb and lower limb muscle strength, balance, and walking speed
Gothe (2016) [91] USA	N=108 Mean age (SD): yoga group=62.1 (5.82); control group=62.0 (5.39) years	Hatha yoga, 3 times a week, 60 min per session, 8 weeks	1. AC: Stretching–strengthening control group	Physical function: Mobility, upper and lower limb strength, upper and lower body flexibility, balance, walking speed
Gothe (2013) [68] USA	N=108 Mean age (SD): yoga group=62.1 (5.82); control group=62.0 (5.39) years	Hatha yoga, 3 times a week, 75 min per session, 8 weeks	1. AC: Stretching–strengthening control group	HRQoL: Stress, anxiety

Study id Country	Participants (number, mean age (SD))	Intervention group (type, frequency, session duration, length of intervention)	Control group(s)	Outcome measures
Haber (1983) [70] USA	Centre1 N: 51 Mean age: 69 years Centre1 N: 35 Mean age: 70 years	The Easy Does it Yoga Programme for Older People, 1 session a week, Daily home practice encouraged, 10 weeks	1. IC: Control group (film series or art class)	HRQoL: Self assessed health status, psychological wellbeing
Haber (1988) [66] USA	*N*=410 Mean age: 75 years	The Easy Does it Yoga Programme for Older People, 3 times a week, 60 min per session, 10 weeks	1. IC: Control group	HRQoL: Self-care, sociability, medication usage
Hariprasad (2013) [74] India	*N*=87 Mean age (SD): yoga group=75.74 (6.46); control group=74.78 (7.35) years	Yoga, daily supervised sessions for 1 month. 1 session per week in the 2nd and 3rd month. Daily home practice following this, 60 min per session, 24 weeks	1. IC: Wait-list control	HRQoL: Perceived physical and mental health, sleep
Krishnamurthy (2007) [92] India	*N*=50 Mean age (SD): yoga group=70.1 (8.3); ayurveda=72.1 (9.0); wait-list=72.3 (7.4) years	Yoga, 6 times a week, 60 min per session, 24 weeks	1. IC: Ayurveda group (herbal preparation) 2. IC: Wait-list control	Physical function: Balance, mobility
Krishnamurthy (2007)2 [58] India	*N*=50 Mean age (SD): yoga group=70.1 (8.3); ayurveda=72.1 (9.0); wait-list=72.3 (7.4) years	Yoga, 6 times a week, 60 min per session, 24 weeks	1. IC: Ayurveda group (herbal preparation) 2. IC: Wait-list control	HRQoL: Depression

Study id Country	Participants (number, mean age (SD))	Intervention group (type, frequency, session duration, length of intervention)	Control group(s)	Outcome measures
Leininger (2006) [69] USA	*N*=82 Mean age (SD): yoga group=69.6 (6.7); education group=68.2 (5.4) years	Hatha yoga, 2 supervised sessions a week. Home exercises recommended at least three times a week, 60 min per session, 10 weeks	1. IC: Education control group (on topics of osteoporosis and fitness)	Physical function: Balance, lower limb strength, HRQoL: Balance confidence, vitality
Manjunath (2005) [57] India	*N*=50 Mean age (SD): yoga group=70.1 (8.3); ayurveda=72.1 (9.0); wait-list=72.3 (7.4) years	Yoga training, 6 times a week, 60 min per session, 24 weeks	1. IC: Ayurveda (herbal preparation) 2. IC: Wait-list control	HRQoL: Sleep quality
Marques (2017) [78] Portugal	*N*=25 Mean age (SD): 83.16 (7.4) years	Chair based yoga, 2 to 3 times a week, 50 min per session, 28 weeks	1. IC: Control group given education booklet	Physical function: Cardio-respiratory fitness, mobility, upper body flexibility HRQoL: Stress, perceived mental health
Morris (2008) [67] USA	*N*=18 Mean age (SD): 76.06 (6.35) years	Hatha yoga, 2 times a week, 60 min per session, 8 weeks	1. AC: Balance training exercise 2. IC: Fall risk awareness	Physical function: Balance, fall frequency HRQoL: Fear of falls, balance confidence
Ni (2014) [56] USA	*N*=39 Mean age (SD): 74.15 (6.99) years	Balance yoga programme, 2 times a week, 60 min per session, 12 weeks	1. AC: Tai Chi 2. AC: Standard balance programme	Physical function: Mobility, balance, walking speed

Study id Country	Participants (number, mean age (SD))	Intervention group (type, frequency, session duration, length of intervention)	Control group(s)	Outcome measures
Nick (2016) [76] Iran	*N*=39 Mean age (SD): yoga group=68 (4.87); control group=68.79 (4.81) years	Hatha yoga, 2 times per week, 60 min per session, 8 weeks	1. IC: Control group	Physical function: Balance HRQoL: Fear of falls
Noradechanunt (2017) [72] Australia	*N*=33 Mean age (SD): 67.7 (6.7) years	Thai Yoga, 2 supervised session a week, 80 min per session. Home practice on alternate days for 20 min, 12 weeks	1. AC: Tai Chi 2. IC: Control group	Physical function: Lower and upper limb strength, lower and upper body flexibility, mobility HRQoL: Perceived physical and mental health, depression
Oken (2006) [64] USA	*N*=118 Mean age (SD): yoga group=71.5 (4.9); exercise group=73.6 (5.1); wait-list=71.2 (4.4) years	Iyengar yoga, 1 class a week with home practice, 90 min per session, 24 weeks	1. AC: Walking group 2. IC: Wait-list control	Physical function: Lower body flexibility, lower limb strength, balance, walking speed HRQoL: Mood, fatigue, depression, perceived physical and mental health, pain, general health and well-being, social functioning, vitality
Saravanakumar (2014) [71] Australia	*N*=33 Mean age (SD): 83.8 (7.9) years	Yoga, 2 times a week, 30 min per session, 14 weeks	1. AC: Tai Chi 2. IC: Usual care	Physical function: Balance, fall incidence, HRQoL: Pain, quality of life.

Study id Country	Participants (number, mean age (SD))	Intervention group (type, frequency, session duration, length of intervention)	Control group(s)	Outcome measures
Tew (2017) [77] UK	*N*=47 Mean age (SD): 74.8 (7.2) years	British Wheel of Yoga Gentle Years Yoga programme, 10 sessions during a 12-week period, 75 min. Home practice encouraged for 10–20 min on most days, 12 weeks	1. IC: Wait-list control	Physical function: Body weight, BMI, waist circumference, lower limb strength, upper and lower body flexibility, balance, walking speed HRQoL: Perceived mental health, quality of life
Tiedemann (2013) [73] Australia	*N*=52 Mean age (SD): 68 years (7.1) years	Iyengar yoga, 2 session a week, 60 min. Home practice 2 days a week for 10–20 min, 12 weeks	1. IC: Control group given fall prevention education booklet	Physical function: Balance, lower limb strength, walking speed HRQoL: Fear of falls
Vogler (2011) [39] Australia	*N*=38 Mean age (SD): 73.21 (8.38) years	Iyengar yoga, 2 times per week, 90 min per session. Home practice 3 days per week for 15–20 min, 8 weeks	1. IC: Wait-list control group	Physical function: Muscle strength, range of motion of the upper extremity, hip flexion, hip extension, hip abduction, and trunk rotation HRQoL: General health and well-being, perceived physical and mental health

Study id Country	Participants (number, mean age (SD))	Intervention group (type, frequency, session duration, length of intervention)	Control group(s)	Outcome measures
Wang (2010) [65] USA	N=18 Mean age (SD): 74.9 (8.4) years	Yoga group, 2 times per week, 60 min per session, 4 weeks	1. IC: Social group	Physical function: Balance, lower limb strength, lower body flexibility HRQoL: Depression, morale, hope, social isolation

Key: HRQoL: health related quality of life; BMI: Body Mass Index; AC: Active control; IC: Inactive control; N: Number of participants analysed in included studies

Table 6.5: Below summarizes the various kinds of yoga practices encompassed in the study (Sivaramakrishnan et al, 2019: *Open Access; This article is distributed under the terms of the Creative Commons Attribution 4.0 International License (http://creativecommons.org/licenses/by/4.0/), which permits unrestricted use, distribution, and reproduction in any medium, provided you give appropriate credit to the original author(s) and the source, provide a link to the Creative Commons license, and indicate if changes were made. The Creative Commons Public Domain Dedication waiver (http://creativecommons.org/publicdomain/zero/1.0/) applies to the data made available in this article, unless otherwise stated*)

Types of yoga in included studies (number of studies, total number of participants)	**Description**
The types of yoga used in studies are similar in structure and postures, and their main features are highlighted below.	
Hatha yoga (4 studies, 247 participants)	Traditional yoga that includes combinations of postures, breathing, and meditation
Chair yoga (3 studies, 165 participants)	This essentially follows a traditional Hatha yoga format, but is modified so that chairs are used during practice to accommodate physical limitations
Iyengar yoga (3 studies, 208 participants)	Created by BKS Iyengar; based on Hatha yoga, but emphasis is on strength, balance, and use of props. Usually involves slow movement and holding poses

Types of yoga in included studies (number of studies, total number of participants)	**Description**
Silver Yoga (2 studies, 231 participants)	The programme is based on Hatha yoga and Raja yoga (type of yoga that focuses on concentration and meditative techniques). The programme includes gentle stretching postures to increase range of motion and progressive muscle relaxation. Special consideration given for the physical abilities and tolerance of older adults
Balance yoga programme (1 study, 39 participants)	This programme is based on a study by the authors showing specific muscle utilization patterns during different flow-based yoga poses. The programme has three levels of difficulty, becoming progressively challenging
The Easy Does It Yoga Programme (2 studies, 496 participants)	Yoga programme designed for older adults
Thai Yoga (1 study, 33 participants)	Thai Yoga is similar to the Hatha yoga style. However, it is less strenuous and incorporates postures that are less challenging and easier to perform than those of Hatha yoga
British Wheel of Yoga (BWY) Gentle Years Yoga programme (1 study, 47 participants)	The British Wheel of Yoga (BWY) Gentle Years Yoga programme was developed to cater to the needs of older people with age-related conditions (osteoarthritis, hypertension, dementia, and sensory impairment). Hatha yoga poses were adapted so that older adults with comorbidities and physical limitations could safely participate

Balance, flexibility of the lower body, and lower limb strength were improved evidently by yoga when compared against inactive controls; additionally, favorable effects were noticed in perceived physical and mental health, depression, and sleep quality against inactive controls while favorable effects were bestowed by yoga when active controls were taken into account. Here inactive controls refer to the use of educational materials or wait-list controls while active controls are inclusive of chair aerobics or walking.

Group by Outcome	Study name	Hedges's g	Lower limit	Upper limit	p-Value
		Statistics for each study			
Depression	Bethany (2005)_Dep	1.228	0.326	2.130	0.008
Depression	Oken (2006)_Dep	0.171	-0.244	0.586	0.420
Depression	Wang (2010)_Dep	0.207	-0.713	1.126	0.659
Depression	Bonura (2014)_Dep	0.461	-0.026	0.948	0.064
Depression	Chen (2010)1_Dep	0.655	0.116	1.195	0.017
Depression	Chen (2009)_Dep	0.802	0.444	1.161	0.000
Depression	Noradechanunt (2017)_Dep	-0.120	-0.865	0.625	0.752
Depression	Krishnamurthy (2007)2 Ay_Dep	1.480	0.537	2.423	0.002
Depression	Krishnamurthy (2007)2 WL_Dep	1.403	0.557	2.248	0.001
Depression		0.635	0.322	0.948	0.000
Fear of falls	Morris (2008)_FOF	-0.185	-1.226	0.857	0.728
Fear of falls	Nick (2016)_FOF	1.197	0.527	1.867	0.000
Fear of falls	Tiedemann (2013)_FOF	0.058	-0.478	0.594	0.833
Fear of falls		0.393	-0.451	1.237	0.362
Per mental health	Vogler (2011)_PMH	1.383	0.687	2.079	0.000
Per mental health	Oken (2006)_PMH	0.085	-0.329	0.500	0.686
Per mental health	Chen (2009)_PMH	0.903	0.541	1.265	0.000
Per mental health	Bonura (2014)_PMH	0.443	-0.043	0.930	0.074
Per mental health	Wang (2010)_PMH	0.210	-0.709	1.130	0.654
Per mental health	Hariprasad (2013)_PMH	0.807	0.437	1.177	0.000
Per mental health	Marques (2017)_PMH	0.067	-0.707	0.841	0.865
Per mental health	Noradechanunt (2017)_PMH	0.539	-0.219	1.298	0.164
Per mental health	Tew (2017)_PMH	0.719	0.136	1.303	0.016
Per mental health		0.597	0.328	0.867	0.000
Per physical health	Vogler (2011)_PPH	0.193	-0.431	0.817	0.544
Per physical health	Oken (2006)_PPH	0.519	0.098	0.940	0.016
Per physical health	Chen (2009)_PPH	0.968	0.603	1.332	0.000
Per physical health	Hariprasad (2013)_PPH	0.912	0.538	1.286	0.000
Per physical health	Noradechanunt (2017)_PPH	0.032	-0.713	0.776	0.934
Per physical health		0.614	0.285	0.943	0.000
Sleep quality	Chen (2009)_Sleep	0.623	0.270	0.976	0.001
Sleep quality	Chen (2010)1_Sleep	1.052	0.491	1.612	0.000
Sleep quality	Hariprasad (2013)_Sleep	0.396	0.036	0.755	0.031
Sleep quality	Manjunath (2005) Ay_Sleep	0.620	-0.231	1.470	0.153
Sleep quality	Manjunath (2005) WL_Sleep	0.997	0.190	1.803	0.015
Sleep quality		0.645	0.408	0.883	0.000
Social health	Hariprasad (2013)_Soc	0.572	0.209	0.935	0.002
Social health	Oken (2006)_Soc	0.157	-0.258	0.572	0.458
Social health	Wang (2010)_Soc	-0.296	-1.218	0.626	0.530
Social health		0.270	-0.147	0.687	0.204
Vitality	Leininger (2006)_Vit	0.248	-0.182	0.679	0.258
Vitality	Oken (2006)_Vit	0.310	-0.107	0.726	0.145
Vitality	Noradechanunt (2017)_Vit	0.518	-0.240	1.276	0.180
Vitality		0.312	0.034	0.591	0.028

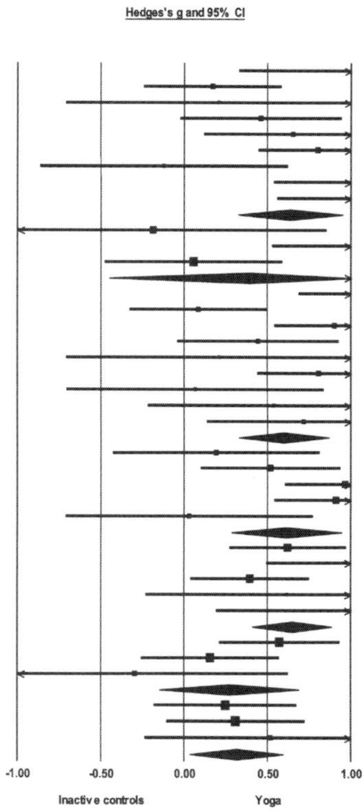

Figure 6.5: Above summarizes the analyses of yoga vs. inactive controls (Sivaramakrishnan et al, 2019: *Open Access; This article is distributed under the terms of the Creative Commons Attribution 4.0 International License (http://creativecommons.org/licenses/by/4.0/), which permits unrestricted use, distribution, and reproduction in any medium, provided you give appropriate credit to the original author(s) and the source, provide a link to the Creative Commons license, and indicate if changes were made. The Creative Commons Public Domain Dedication waiver (http://creativecommons.org/publicdomain/zero/1.0/) applies to the data made available in this article, unless otherwise stated*).

Group by Outcome	Study name	Statistics for each study			
		Hedges's g	Lower limit	Upper limit	p-Value
Anxiety	Bonura (2014)_Anx	0.339	-0.142	0.819	0.167
Anxiety	Bethany (2005) CA_Anx	1.097	0.084	2.111	0.034
Anxiety	Bethany (2005) W_Anx	1.041	-0.036	2.118	0.058
Anxiety	Gothe (2013)_Anx	0.043	-0.333	0.419	0.822
Anxiety		0.426	-0.025	0.876	0.064
Depression	Bethany (2005) CA_Dep	1.317	0.274	2.360	0.013
Depression	Bethany (2005) W_Dep	0.791	-0.259	1.840	0.140
Depression	Oken (2006)_Dep	0.545	0.129	0.960	0.010
Depression	Bonura (2014)_Dep	0.526	0.041	1.012	0.033
Depression	Noradechanunt (2017)_Dep	0.000	-0.744	0.744	1.000
Depression		0.538	0.247	0.829	0.000
Per mental health	Oken (2006)_PMH	0.320	-0.090	0.731	0.126
Per mental health	Noradechanunt (2017)_PMH	0.366	-0.385	1.117	0.340
Per mental health	Bonura (2014)_PMH	0.133	-0.345	0.610	0.586
Per mental health		0.259	-0.029	0.546	0.078

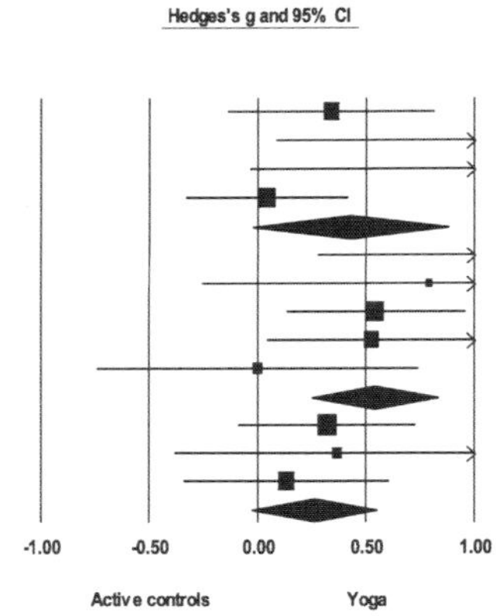

Figure 6.6: Above summarizes the analyses of yoga vs. active controls (Sivaramakrishnan et al, 2019: *Open Access; This article is distributed under the terms of the Creative Commons Attribution 4.0 International License (http://creativecommons.org/licenses/by/4.0/), which permits unrestricted use, distribution, and reproduction in any medium, provided you give appropriate credit to the original author(s) and the source, provide a link to the Creative Commons license, and indicate if changes were made. The Creative Commons Public Domain Dedication waiver (http://creativecommons.org/publicdomain/zero/1.0/) applies to the data made available in this article, unless otherwise stated*).

In a nutshell, this study presents effects described as "small to moderate significant effects" of yoga on body flexibility along with vitality, perceived mental health, and improving depression in elderly individuals when compared to inactive people while even those who were physically active showed "significant small to moderate effects" on depression along with body flexibility due to yoga.

To quote the researchers, "*Yoga is a multimodal activity that improves muscle strength, balance and flexibility in older adults, and physical activity policy should continue to promote yoga as an activity that enhances physical and mental wellbeing in this population*" (Sivaramakrishnan et al, 2019).

6.5 BREATHING AND PRANAYAMA FOR DEPRESSION

Pranayama refers to the terms Prana+ayama or expulsion of vital energy account for the breath pattern aspect of yoga. There are 3 aspects of pranayama: Poorak', 'Kumbhak', & 'Rachek' or the inhalation, exhalation, and the holding of the breath. These can adopt several patterns ranging from nadishodhan, anulomvilom, kapalbhati, bhramri, bhastrika, agnisar, ujjayi, and sheetli .Given these concepts, a research article published in 2017 explored the effect of pranayama on depression. Researchers Jain and Sharma tested the effects of the pranayama techniques: “bhastrika , Kapalbhati , Bahya Pranayama(with tribandh), Anulom-Vilom, Bhramri, Udgeet ‘Omkar Japa’, Pranav Pranayama (Meditation of Om with Breathing)” on 60 individuals over 3 months while another 60 participants served as the controls. The moods of the participants were assessed employing the Eight State Questionnaire (8SQ: Anxiety, Stress, Depression, Regression, Fatigue, Guilt, Extraversion & Arousal).

Table 6.5: Comparison of the effect of the yogic technique on anxiety (AN) and depression (DEP) before treatment (AN1/DEP1), after one month, after two months, and three months (AN2/DEP2, AN3/DEP3, and AN4/DEP4) are displayed in Table 6.5 below (Jain and Sharma, 2017*; This is an Open Access Research distributed under the terms of the Creative Commons Attribution License (www.creativecommons.org/licenses/by/2.0), which permits unrestricted use, distribution, and reproduction in any Medium, provided the original work is properly cited*)

Table -2 Mean, SDs and't' value on Anxiety by Experimental and Control group.

VARIABLE		EXPERIMENTAL GROUP [N=60] MEAN	S.D.	CONTROL GROUP [N=60] MEAN	SD	df	t-VALUE
ANXIETY	AN1	19.97	3.01	19.55	3.11	118	.75 (N.S.)
	AN2	15.50	2.18	19.13	2.99	118	7.60***
	AN3	9.25	2.49	19.00	3.32	118	18.17***
	AN4	2.63	1.57	19.65	3.09	118	38.01***

***significant at .001 level
**significant at .01 level
*significant at .05 level

Table -3 Mean, SDs and 't' value on Depression by Experimental and Control group.

VARIABLE		EXPERIMENTAL GROUP [N=60] MEAN	S.D.	CONTROL GROUP [N=60] MEAN	SD	df	t-VALUE
DEPRESSION	DE1	18.58	2.31	18.38	2.10	118	.49 (N.S.)
	DE2	14.65	1.90	18.11	3.09	118	7.40***
	DE3	11.20	2.01	19.53	2.56	118	19.78***
	DE4	7.26	1.63	17.85	1.51	118	36.76***

***significant at .001 level
**significant at .01 level
*significant at .05 level

Figure [1] showing the mean scores on anxiety

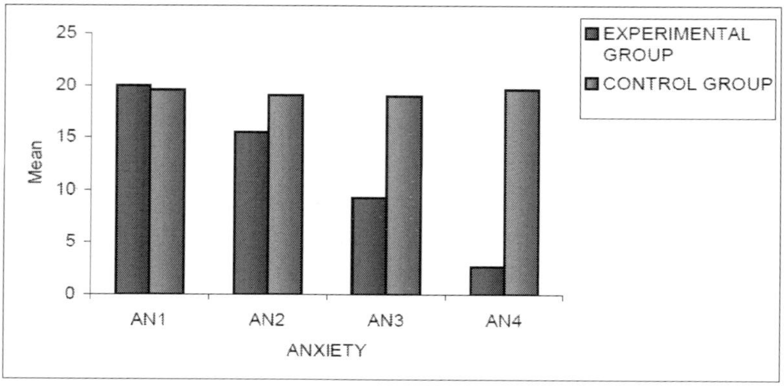

Figure [2] showing the mean scores on depression.

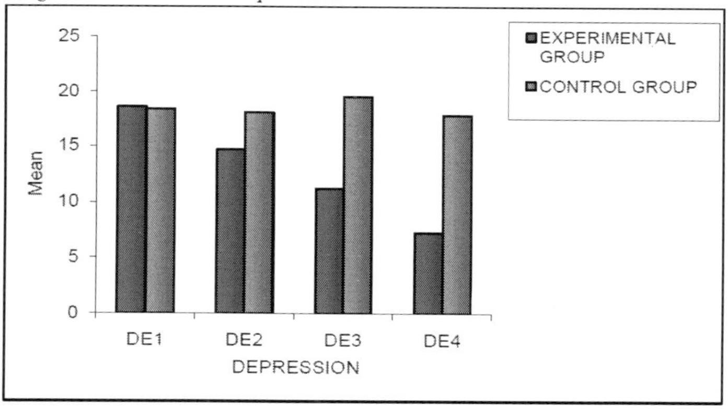

Figure 6.7: Above is representative of the positive effects of pranayama in lowering anxiety and depression when compared with controls (Jain and Sharma, 2017*; This is an Open Access Research distributed under the terms of the Creative Commons Attribution License (www.creativecommons.org/licenses/by/2.0), which permits unrestricted use, distribution, and reproduction in any Medium, provided the original work is properly cited*).

Presenting here the words of the researchers: "*Thus the present study on the one hand helps to confirm the previous results starting the positive effect of Yoga: Pranayama on Anxiety & depression and on the other hand it suggests that one can practice Yoga: Pranayama in the present day society without deviating from the basic principles. These make an individual think and live in a positive way with a philosophical view. This helps him to address the life challenges without distress. But whether we use yoga for spiritual transformation or for the management of Anxiety & depression and physical or social well-being, the benefits are multiple and amazing, if practiced regularly and with faith. From the above discussion it may be concluded that there is a positive effect of Yogic Intervention: Pranayama on Anxiety & depression. It may also be said that Yoga: Pranayama provides us perfect physical, social and psychological well-being even under stressful conditions*".

6.6 HATHA YOGA FOR DEPRESSION

Among the various forms of yoga that are based on the *Yoga Sutras, 8 steps describe yoga: ethics,* self-discipline, body postures, regulating the breath, control of senses, and three meditative practices that get progressively deeper. One branch of yoga is hatha yoga based on diet, breathing, and body postures.

A critical review of the effects of hatha yoga on depression was presented by Uebelacker and team (2010) on the basis of literature review and analyses. 5 of the 8 studies were based on clinical depression while 3 were for increased depressive symptoms. Against the backdrop of the inadequacy of current approaches to target depression, the researchers stated that "*Yoga may be an attractive alternative to or a good way to augment current depression treatment strategies*".

Researchers Balasubramaniam and team (2013) assessed the effect of yoga on neuropsychiatric issues by querying the Cochrane Central Register of Controlled Trials and the standard bibliographic databases, MEDLINE, EMBASE, and PsycINFO (April 2011). The study was reported as "the first review to systematically examine the published literature on benefits of yoga for several major psychiatric illnesses".

Of 124 trials obtained, the criteria set by the Research and Development/ University of California at Los Angeles (RAND/UCLA) Appropriateness Method as Grades A (recommended), B (suggested), or C (may be considered) was analyzed.

The authors finalized 4 randomized clinical trials assessing the effects of yoga on depression. It was concluded that"*a potential acute benefit for yoga exists in depression*" (Grade B).

Term	Level	Evidence levels	Explanation
Recommended	A	1 **or 2**	Assessment supported by a substantial amount of high quality (levels 1 or 2) evidence and/or based on consensus of clinical judgment
Suggested	B	1 or 2 - few studies 3 or 4 - many studies and expert consensus	Assessment supported by sparse high grade (Level 1 or 2) data or a substantial amount of low grade (level 3 or 4) data and/or clinical consensus
May be considered	C	3 or 4	Assessment is supported by low grade data without the volume to recommend more highly and likely subject to revision with further studies

Study	Sample	Treatment groups	Intervention	Duration	Outcome measurements	Findings	RCT	Evidence level
Shahidi el al. (2011)	70 depressed women aged 60-80 years from a cultural community in Iran with Geriatric Depression Scale score > 10	Laughter yoga (n = 23), exercise therapy (n=23). and wait-list control groups (n =24)	Laughter yoga consisted of brief talk about something delightful, clapping hands, simple chants simulating diaphragmatic breathing, Gibberish sounds. Combines yoga, breathing, and stretching techniques	10 sessions	Yesavage Geriatric Depression Scale and Diener Life Satisfaction Scale (LSS)	Significant improvement in GDS scores in both laughter and exercise groups compared to controls but not when compared to each other	13 (not double blinded, funding information not given)	2 (Low quality RCT due to insufficient follow-up)

Krishna-murthy and Telles (2007)	69 participants (males and females), older than 60, living in a residential home	Stratified sampling and random allocation to yoga, ayurveda. wait-1st control groups	Yoga consisted of 7 h 30 min weekly sessions of physical postures, relaxation techniques, regulated breathing, devotional songs, and lectures	24 weeks	Shortened version of Geriatric Depression Scale (GDS)	The yoga group showed significant decrease in depression at 3 and 6 months compared to the ayurveda group	13 (not double blinded, funding information not given)	2 (Low quality RCT due to <80% follow-up rate)
Vedamu-rthachar et al. (2006)	Mates aged 18-55 years with alcohol dependence, admitted for the first time to the de-addiction center of NIMHANS, not having serious medical illnesses, schizophrenia, or mania	SKY - Sudarksha Kriya yoga (n=30) therapy, continued inpatient care (n = 30)	SKY consisted of practice of three distinct breathing patterns	2 weeks	BDI scores, ACTH, and cortisol levels	Statistically significant decrease in BDI scores in the SKY group compared to controls. Greater reduction in serum cortisol and ACTH levels in the SKY group	15 (not double blinded)	2 (Low quality RCT due to insufficient follow-up)
Woolery et al. (2004)	28 volunteers aged 18-29 years, with self-reported symptoms of depression, but not on psyc-hotropic treatment and without previous exposure to yoga	Yoga (n 13), wait-list control (n= 15)	1 h weekly lyengar yoga classes, consisting of training in yoga postures	5 weeks	BDI, State-Trait Anxiety Inventory, Profile of mood states, morning cortisol levels	Statistically significant decrease in BD scores, anxiety scores, and higher morning cortisol levels in the yoga group	13 (not double blinded, funding information not given)	2 (Low quality RCT due to insufficient follow-up)

Figure 6.8: Above summarizes the studies assessed for the promising effect of yoga on depression in the systemic review (Balasubramaniam, Telles and Doraiswamy, 2013: *This is an open-access article distributed under the terms of the Creative Commons Attribution License, which permits use, distribution*

and reproduction in other forums, provided the original authors and source are credited and subject to any copyright notices concerning any third-party graphics etc).

In a 2018-published study, researchers Vollbehr and team performed a meta-analysis to look at 2 aspects:

"*(1) is hatha yoga an effective treatment for acute mood and anxiety disorders, and*

(2) is hatha yoga an effective treatment for chronic and/or treatment-resistant mood and anxiety disorders?"

The approach involved looking at randomized controlled trials with patients with mood and anxiety disorders across Medline, Cochrane Library, Current Controlled Trials, Clinical Trials.gov, NHR

Centre for Reviews and Dissemination, PsycINFO, and CINAHL. Following the addressing of publication bias via funnel plots, the symptoms of mood were assessed as the outcome. 18 trials were selected with a total of 1,532 participants (85 mean participants; average age: 36.6 years) from theUS, India, Sweden, and Germany.

Study	Total patients (yoga group), diagnosis	Diagnosis by	Co-interventions	Intervention groups, length, frequency, duration, amount of home practice		Length of intervention, follow-up	Outcome measures 1. Primary 2. Secondary	Results		CTAM
				Treatment	Control			Short term	Long term	
Broota & Dhir, 1990 [91]	30 (10*) adults with neurotic/ reactive depression (baseline scores unknown), outpatients from a psychiatry department	Psychiatrist	Antidepressant medication	Broota Relaxation Technique (BRT, postures, breathing), 1x 20 min/day, home practice not reported	1. PMR, 1x 20 min/ day 2. Control group, narrating about problems, 1x 20 min/ day	3 days	1. Symptoms of depression (Symptom check list)	1. Sig dif favoring BRT over control 2. BRT slightly superior to JPR		36
Falsafi, 2016 [88]	90 (30*) students with depression and/or anxiety disorder (baseline BDI 20.0/21.1/20.2; HAM-A 21.2/22.2/21.1), recruitment at the university	Health care professio-nal	Regular treatment (medication and/or psychotherapy)	Hatha Yoga (postures, meditation), 1x 75 min/week, daily 20 min home practice	1. Mindfulness training, 1x 75 min/ week, daily 20 min home practice 2. Control group (no intervention)	8 weeks, 12 weeks	1. Symptoms of depression (BDI) 2. Symptoms of anxiety (HAM-A)	1. Sig dif favoring yoga and mindfulness over control, no sig dif between yoga and mindfulness 2. Sig dif favoring yoga and mindfulness over control, no sig dif between yoga and mindfulness	1. Sig dif favoring yoga and mindfulness over control, no sig dif between yoga and mindfulness 2. Sig dif favoring yoga and mindfulness over control, no sig dif between yoga and mindfulness	37

Field et al., 2013 [79]	92 (46*) prenatally depressed women (MDD, dysthymia) (baseline CES-D 33.0/35.1; STAI 53.4/55.0), community sample	SCID	None	Yoga (postures), 1x 20 min/ week, home practice with manual and dvd, not reported how frequent	Social support group, 1x 20 min/week	12 weeks, postpartum (1–3 weeks post birth)	1. Symptoms of depression (CES-D, EPDS) 2. Symptoms of anxiety (STAI)	1. No sig group dif 2. No sig group dif	1. No sig group dif 2. No sig group dif	67
Field et al., 2012 [80]	84 (unclear*) prenatally depressed women (MDD, dysthymia) (baseline CES-D 28.35/24.08/22.65; STAI 50.0/44.19/ 42.38), community sample	SCID	Mostly none (95%)	Yoga (postures), 1x 20 min/week, home practice not reported	1. Massage, 1x 20 min/ week 2. Standard prenatal care	12 weeks	1. Symptoms of depression (CES-D) 2. Symptoms of anxiety (STAI)	1. Sig dif favoring yoga over control group, no sig dif between yoga and massage 2. Sig dif favoring yoga over control group, no sig dif between yoga and massage		64

Helgadót-tir et al., 2016 & 2018 [93, 97]	620 (106*) adults with a depressive disorder or anxiety disorder (baseline MADRS 21.5), community sample	MINI	Standard treatment for depression	Yoga (postures), 3x 55 min/week	1. Intermediate-level aerobics, 3x 55 min/week 2. More strenuous aerobics, 3x 55 min/week 3. Treatment as usual	12 weeks, 1 year	1. Symptoms of depression (MADRS)	1. Sig dif favoring yoga group over control, trend towards a sig diff favoring yoga over moderate exercise group, no sig diff between yoga and vigorous exercise group	1. Sig dif favoring yoga group over control, sig dif favoring yoga over moderate exercise, no sig dif between yoga and vigorous exercise group	68
Kinser et al., 2013 & 2014 [86, 34]	27 (15*) women with MDD or dysthymia (baseline PHQ-9 14.8/18.3; STAI 52.5/55.1), community sample	MINI	Usual depression care	Gentle Hatha yoga (postures, breathing, relaxation), 1x 75 min session/ [90](89)(89) week, daily home practice with dvd	Health education sessions, 1x 75 min/ week, weekly review handout at home	8 weeks, 1 year	1. Symptoms of depression (PHQ-9) 2. Symptoms of anxiety (STAI)	1. No sig group dif 2. No sig group dif	1. Sig dif favoring yoga over control 2. No sig group dif	40
Mitchell et al., 2012 [81]	24 (12*) prenatally depressed women (MDD) (baseline CES-D 22.42/27.5), community sample	SCID	Not reported	Yoga (postures), 2x 20 min/ week, home practice not reported	Parenting education, 2x 20 min/ week	12 weeks	1. Symptoms of depression (CES-D)	1. Sig dif favoring yoga group over control		39

Parthasa-rathy et al., 2014 [78]	45 (unclear*) women with anxiety disorder (baseline TAS 114.4/114.46/ 114.27), from a tertiary care center	Unclear	Not reported	Yoga (postures, breathing, relaxation), 45 min/day	1. Integrated yoga module (more postures, breathing, relaxation), 45 min/day 2. Control group with no special activities	8 weeks	1. Symptoms of anxiety (TAS)	1. Reduction in both yoga groups, not control group. Sig dif favoring integrated yoga module over yoga		32
Prathi-kanti et al., 2017 [89]	38 (20*) adults with major depression (baseline BDI 22.4), community sample	MINI	1 participant took psychotherapy	Hatha yoga (postures, breathing, meditation), 2x 90 min / week, home practice not reported	Education modules on yoga history and philosophy, 16x 90 min / week, home practice not reported	8 weeks	1. Symptoms of depression (BDI)	1. Sig dif favoring yoga group over control		49
Sahasi et al., 1989 [84]	91 (38*) adults with anxiety neurosis (baseline scores unknown), outpatients from a psychiatric center	Psychiatrist	Not reported	Yoga (postures, breathing, meditation), 7x 40 min session/ week (5x with instructor, 2x at home)	Diazepam (no dose or frequency given)	3 months	1. Symptoms of anxiety (IPAT)	1. Sig reduction for yoga group, not for control group		34
Sarubin et al., 2014 [94]	60 (22*) adults with MDD (baseline HAM-D 22.04), unclear recruitment	SCID	Quetiapine or escitalopram	Hatha yoga (no description given), 1x 60 min / week, home practice not reported	Control group (no yoga)	5 weeks	1. Symptoms of depression (HAM-D)	1. No sig group dif		35

Schuver & Lewis, 2016 [35]	40 (20*) depressed women (baseline BDI 22.36), community sample	SCID	Usual depression care	Mindfulness-based yoga (postures, breathing, meditation), by a DVD, 2x 60–75 min / week	Walking, with a DVD, 2x 60 min / week	12 weeks, 1 month	1. Symptoms of depression (BDI)	1. No sig group dif	1. No sig group dif	37
Tolahunase et al., 2018 [92]	58 (29*) adults with MDD (baseline BDI 26.96/28.10), outpatients from a psychiatry department	Unclear	Routine drug treatment for at least 6 months	Yoga- and meditation-based lifestyle interventions (postures, breathing, meditation, lectures on lifestyle), 5x 120 min / week	Control group (no yoga)	12 weeks	1. Symptoms of depression (BDI)	1. Sig dif favoring yoga group over control		52
Uebelacker et al., 2016 [82]	20 (12*) pregnant women with depression (major or minor depression) (baseline QIDS 11.17/11.5), community sample	SCID	1 person took antidepressant medication	Gentle yoga (postures, breathing, meditation), 1x 75 min / week, home practice encouraged	Mom-baby wellness workshop, 1x 75 min / week	9 weeks	1. Symptoms of depression (QIDS, EPDS)	1. No sig group dif		42

Note. BAI, Beck Anxiety Inventory; BDI, Beck Depression Inventory; CBT, Cognitive Behavioral Therapy; CES-D, Center for Epidemiological Studies Depression Scale; CIS, Clinical Interview Schedule; dif, difference; EPDS, Edinburgh Postnatal Depression Scale; fu, follow-up; GAD, Generalized Anxiety Disorder; HAM-A, Hamilton Rating Scale for Anxiety; HAM-D, Hamilton Rating Scale for Depression; IPAT, Institute for Personality and Ability Testing [anxiety scale]; LoE, Level of Evidence; MADRS, Montgomery-Asberg Depression Rating Scale; MDD, Major Depressive Disorder; min, minutes; MINI, MINI International Neuropsychiatric Interview; PHQ, Patient Health Questionnaire; PMR, Progressive Muscle Relaxation; QIDS, Quick Inventory of Depressive Symptomatology; SCID, Structured Clinical Interview for DSM-IV; sig, significant; STAI, State-Trait Anxiety Questionnaire; TAS, Taylor's Anxiety Scale;

* Number of patients that received the yoga intervention.

Study	Total patients [yoga group], diagnosis	Diagnosis by	Co-interventions	Intervention groups, length, frequency, duration, amount of home practice		Length of intervention, follow-up	Outcome measures 1. Primary 2. Secondary	Results		CTAM
				Treatment	Control			Short term	Long term	
Butler et al., 2008 [87]	52 (17*) adults with dysthymia, double depression, (chronic) MDD (baseline HAM-D 15.87/12.33/ 15.81), community sample	SCID	Medication, no psychotherapy	Meditation and Hatha yoga (postures, breathing, meditation), 1x 2hr/week, daily home practice encouraged with audiocassettes (30 min/day, 6x /week)	1. Group therapy with hypnosis, 1x 1.5 hr/week 2. Control group, psychoeducation	12 weeks, 9 months	1. Symptoms of depression (HAM-D; CDRS-SR) 2. MDE; Remission > 2 months	1. No sig group dif	1. No sig group dif 2. MDE: sig dif favoring both groups over control; Remission: sig dif favoring yoga over control	60
Gupta et al., 2013 [83]	12 (6*) adults with GAD (baseline HAM-A 30.83/32.0), outpatients from institute for yoga and naturopathy	Unclear	Not reported	Yoga (postures, breathing), 1 hr/ day, home practice not reported	Naturopathy, 2x 30 min/day	3 weeks	1. Symptoms of anxiety (HAM-A)	1. No sig group dif, more improvement in yoga group		29

Uebela-cker et al., 2017 [90]	122 (63*) adults with major depressive disorder (baseline QIDS 12.87), community sample (2/3 chronic)	SCID	Antidepressant medication (95–100%), psychotherapy (40%)	Hatha yoga (postures, breathing, meditation), at least 1 x 80 min / week, home practice encouraged	Healthy living workshop, 1–2 x 60 min / week	10 weeks, 3 and 6 months	1. Symptoms of depression (QIDS, PHQ-9)	1. No sig group dif	3 months: sig dif favoring yoga group over control 6 months: sig dif favoring yoga group over control	82
Vahia et al., 1973, stage 2 [85]	27 (15*) adults with psychoneur-osis, absence of response to conventional treatments (baseline TAS 25.53/29.83), outpatients from a hospital	Unclear	Placebo tablet, support, reassurance	Yoga (postures, breathing, meditation), 1 hr, 5 days / week	Relaxation resembling yoga, 1 hr, 5 days / week	4 weeks (at least)	1. Symptoms of anxiety (TAS)	1. Sig more improvement in yoga than control		56

Note 1. CBT, Cognitive Behavioral Therapy; CDRS-SR, Cornell Dysthymia Rating Scale; dif, difference; HAM-A, Hamilton Rating Scale for Anxiety; HAM-D, Hamilton Rating Scale for Depression; LoE, Level of Evidence; MDE, Major Depressive Episode; min, minutes; MINI, MINI International Neuropsychiatric Interview; QIDS, Quick Inventory of Depressive Symptomatology; SCID, Structured Clinical Interview for DSM-IV; sig, significant; TOP, Treatment Outcome Package; sign, significant; TAS, Taylor's Anxiety Scale;

* Number of patients that received the yoga intervention.

Figure 6.9: Above displays ***TOP***: acute population features; ***BOTTOM***: chronic/treatment-resistant populations (Vollbehr et al, 2018: *This is an open-access article distributed under the terms of the Creative Commons Attribution License, which permits unrestricted use, distribution, and reproduction in any medium, provided the original author and source are credited*).

Of the 18 studies, 4 involved chronic patients while 14 involved acute ones. When compared against psychoeducation controls, there was an evident lowering of depression due to Hatha yoga: Cohen's *d* -0.52 (95% *CI* = -0.96, -0.08). Overall, the researchers reported that chronic depression can benefit from hatha yoga practices.

A team of scientists conducted what they described as "*a prospective, single-center, single-blind, randomized, controlled, parallel group pilot trial of an 8-week hatha yoga program as mono-therapy in major depression*" and published the work in 2017 (ClinicalTrials.gov NCT01210651).

The study involved the recruitment of 38 adults from the San Francisco community with mild-to-moderate major depression, as assessed by the Mini International Neuropsychiatric Interview and scores of 14–28 on the Beck Depression Inventory-II (BDI).

The initial screening involved eliminating those people who were on mind-body practices, antidepressants, psychotherapy, or herbal or nutraceutical-based approaches. Random assigning of the participants was done into a control group who received yoga history modules twice a week for 8 weeks and the other received hatha yoga exercises twice a week for 8 weeks (both were led by instructors).

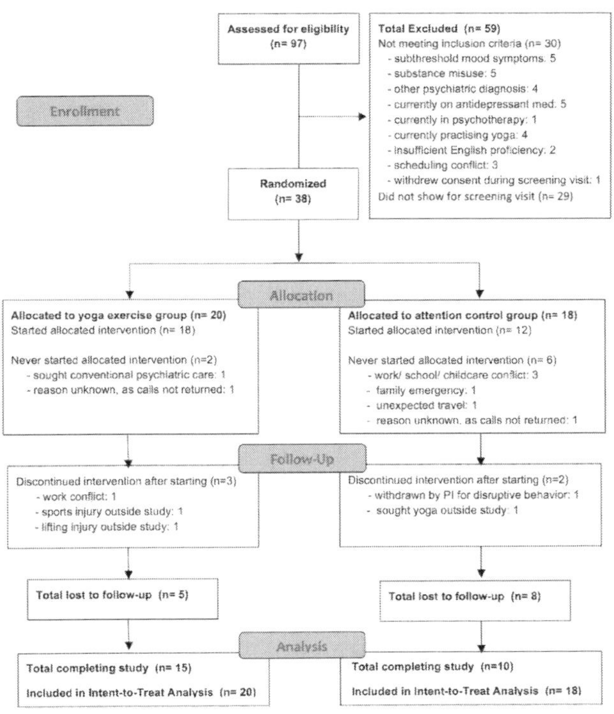

Covariate	Covariate Sub-category	Full Sample, N = 38	Yoga Group, N = 20	Control Group, N = 18	P-value
Age, in Years: Mean (SD), Range		43.4 (14.8), 22–72	43.1 (15.2), 22–64	43.8 (14.7), 23–72	0.89
Female Gender		26 (68%)	15 (75%)	11 (61%)	0.49
Married		4 (11%)	1 (5%)	3 (17%)	0.33
Ethnicity*					0.77
	European descent	22 (58%)	13 (65%)	9 (50%)	
	Asian descent	6 (16%)	2 (10%)	4 (22%)	
	Latino descent	4 (11%)	2 (10%)	2 (11%)	
	African descent	3 (8%)	1 (5%)	2 (11%)	
	Multi-ethnic	3 (8%)	2 (10%)	1 (6%)	
Highest Education					0.19
	High school diploma	2 (5%)	0 (0%)	2 (11%)	
	Two years of college	12 (32%)	5 (25%)	7 (39%)	
	Four-year college degree or higher	24 (63%)	15 (75%)	9 (50%)	
Occupational Status*					0.29
	Employed	20 (53%)	8 (40%)	12 (67%)	
	Unemployed	9 (24%)	7 (35%)	2 (11%)	
	Student	4 (11%)	2 (10%)	2 (11%)	
	Retired	5 (13%)	3 (15%)	2 (11%)	
Previous Yoga Exposure		24 (64%)	10 (50%)	14 (78%)	0.10
Number of Previous Depressive Episodes*					0.74
	None	4 (11%)	3 (15%)	1 (6%)	
	1 Episode	22 (58%)	12 (60%)	10 (56%)	
	2 Episodes	9 (24%)	4 (20%)	5 (28%)	
	3 Episodes	3 (8%)	1 (5%)	2 (11%)	
Number of Previous Antidepressant Trials*					0.24
	None	13 (34%)	10 (50%)	3 (17%)	
	1 Trial	15 (39%)	5 (25%)	10 (56%)	
	2 Trials	5 (13%)	2 (10%)	3 (17%)	
	3 Trials	4 (11%)	3 (15%)	1 (6%)	
	4+ Trials	1 (3%)	0 (0%)	1 (6%)	
Screening BDI Score: Mean (SD)		22.4 (4.5)	22.8 (4.4)	22.4 (4.6)	0.81
	Score 14–19 (Mild depression)	8 (21%)	4 (20%)	4 (22%)	
	Score 20–28 (Moderate depression)	30 (79%)	16 (80%)	14 (78%)	
Screening MMSE Score: Mean (SD)		28.0 (1.6)	27.7 (1.8)	28.4 (1.2)	0.15

BDI = Beck Depression Inventory-II-II; MMSE = Folstein Mini Mental Status Exam
* Percentage total of sub-categories may not equal 100% when summed, due to rounding.

https://doi.org/10.1371/journal.pone.0173869.t002

Figure 6.10: Above shows the system adopted to assess the effect of hatha yoga on depressed individuals. ***TOP***: Flow adopted; ***BOTTOM***: Participant details; moderate depression was seen in 79% of the participants (Prathikanti et al, 2017: *This is an open-access article, free of all copyright, and may be freely reproduced, distributed, transmitted, modified, built upon, or otherwise used by anyone for any lawful purpose. The work is made available under the Creative Commons CC0* public domain *dedication*).

This assigning of the participants was such that the numbers of those with mild depression (BDI scores of 14–19) and those with moderate depression (BDI scores of 20–28) were the same in both groups.

The hatha yoga administered involved mindful body postures (a chest-opening element included believed to address depression), breathing (pranayama), and final relaxation for 90 minutes as led by a certified yoga teacher with a registered and licensed nurse. Apart from the use of props for support, the yoga practices were according to the comfort and capacity of each participant.

Segment	Description
Introduction & Guidelines (5 mins)	Instructor welcomes all participants, including those joining session for first time. Reviews guidelines for yoga practice, encouraging participants to remain within their range of motion or comfort in performing any exercise. Emphasizes that accommodations are available for each exercise, so that those with limitations may still take part. Each participant settles into a comfortable seated position with erect spine and neck, either on chair or cross-legged on the ground.
Breath Regulation (20 mins)	*Ujjayi* (Victorious Breath): deep, slow breathing completed at rate of about 4 respirations/minute against mild airway resistance. Participants learn techniques to slightly contract laryngeal muscles and partially close the glottis to maintain mild airway resistance during inspiration and expiration.
	Bhastrika (Bellows Breath): breath is forcefully inhaled and then exhaled through nostrils at a rate of about 20–30 respirations/minute while using strong abdominal muscle contractions.
	Nadi Shodana (Alternate Nostril Breathing): relaxed breathing pattern with inhalation and exhalation done through just one nostril at a time. Breathing occurs at a normal rate of 12 respirations/minute.
	Brahmari (Bumblebee Breath): Slow breathing is done while eyes and ears are covered via specific placement of hands over face and external ear flaps. There is long slow inhalation and then a gentle humming sound is made with exhalation. Breathing occurs at a rate of about 6 respirations/minute.
Mindful Poses & Movement (50 mins)	*Ardhakati Chakrasana* (Half Waist Wheel Pose): Participants stand upright and then gently bend from waist sideways to the right as far as comfortable, while stretching left arm upward. Returning to an upright stance, they repeat exercise, but bending to the left while stretching right arm up.
	Ardha Chakrasana (Half Wheel Pose): Participants stand upright with hands supporting the lower back. They gently bend backwards from the lower lumbar region as far as comfortable, allowing neck to drop slowly while hands continue to support the lower back.
	Pada Hastasana (Hands to Feet Pose): Participants stand upright with arms raised, and gently bend forward until torso is parallel to ground; participants flexible enough continue to bend forward until hands are touching the feet and the head is touching the knees.
	Bhujungasana (Cobra Pose): Participants lie on ground in prone position. They gently arch the head and spine backwards as far as comfortable, supporting the torso with arms bent at elbow and palms firmly on ground.

	Dhanurasana (Bow Pose): Participants lie on the ground in a prone position. Then, knees are bent with feet moving toward buttocks as far as possible, and the neck gently arched backward. Participants flexible enough gently grasp the ankles with their hands.
	Sarvangasana (Shoulder Stand): Participants lie on the ground in a supine position. Participants with sufficient strength and flexibility raise legs slowly until the legs are perpendicular to the ground; other participants position their legs against a wall so that the legs are as perpendicular to the ground as possible. Those with sufficient strength and flexibility then raise the buttocks and trunk off the ground, keeping the shoulders and elbows firmly on the ground and supporting the back with both palms while the legs remain perpendicular to the ground.
	Matsyasana (Fish Pose): Participants lie on ground in supine position. Hands are placed palm down underneath the back, with fingers pointing toward spine. Pressing down on palms, the chest is slowly raised and back is arched as far as comfortable, with weight of torso supported on elbows.
	Setu Bandhasana (Bridge Pose): Participants lie on ground in a supine position. Then the knees are bent while keeping the soles of the feet on the ground. Arms are kept alongside the torso, but with palms on the floor. The hips are lifted off the floor towards the ceiling as far as comfortable, keeping the feet and palms flat on the floor.
	Balasana (Child Pose): Participants kneel on ground with feet tucked under the buttocks, if possible. Participants then bend forward at waist, bringing chest as close to knees as possible, and resting forehead on the ground, if flexible enough. Arms are extended on the ground, palms down.
	Vakrasana (Twisted Pose): Participants sit on ground with both legs outstretched. Then the left leg is bent at knee with sole of the left foot placed on ground alongside the right knee. The upper part of the body is twisted to the left as much as possible, keeping shoulders in plane with the chest and supporting the torso on the left arm, with left hand firmly planted palm-down on the ground. Elbow of the right arm should be straight and pressed against the lateral surface of the bent left knee. After holding this pose, participants return torso, arms and legs to the beginning position. Then participants repeat pose so that right leg is bent and upper body twists to the right side.
	Ustrasana (Camel Pose): Participants kneel on ground, keeping neck and spine perpendicular to ground. Then spine is arched back gently so that chest puffs out. Those flexible enough reach hands back one at a time to grasp heels. Others reach back to blocks placed on either side of feet.

Final Deep Relaxation (15 mins)	*Shavasana* (Corpse Pose): Participants lie on ground in a supine position, with eyes closed. Arms are stretched away from the torso, palms up. Legs are separated about a shoulder width apart, and the feet rotate out slightly. With calm, soft verbalizations, the instructor guides participants into deep relaxation in this pose, as the breath gradually settles into a slow, restful rhythm.

* Hatha yoga exercises adapted from: Nagarathana, R, Nagendra, HR. Yoga Practices for Anxiety and Depression. Bangalore: Swami Vivekananda Yoga Prakashana; 2001.

Figure 6.11: Above shows the scheme of the hatha yoga elements to address depression in this clinical trial (Prathikanti et al, 2017: *This is an open-access article, free of all copyright, and may be freely reproduced, distributed, transmitted, modified, built upon, or otherwise used by anyone for any lawful purpose. The work is made available under the Creative Commons CC0* public domain *dedication*).

The analyses of the primary outcome: BDI score assessment every 2 weeks were conducted by blinded assessors. The General Self-Efficacy Scale (GSES) and the Rosenberg Self-Esteem Scale (RSES) were employed to also assess self-efficacy and self-esteem. Analyses revealed that the BDI scores of the participants of the yoga program were conspicuously lower than the control. The yoga group displayed a dip in the adjusted mean BDI score by 9.47 points [95% CI, 12.37 to 6.57] while the values were lowered by 2.99 points [95% CI, 6.43 to 0.45] in the control from baseline to the end of the study (8 weeks).

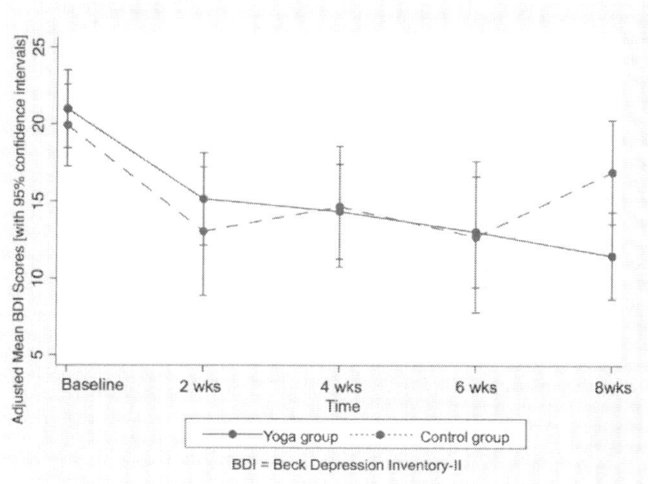

Time	Adjusted Mean BDI Score [95% CI]		8-wk BDI Change Score [95% CI]	
	Yoga Practice Group, n = 20	Attention Control Group, n = 18	Yoga Practice Group, n = 20	Attention Control Group, n = 18
Baseline	20.98 [18.45, 23.51]	19.92 [17.26, 22.58]	-9.47 [-12.37, -6.57]	- 2.99 [-6.43, -0.45]
2 wks	15.15 [12.17, 18.12]	13.06 [8.91, 17.20]		
4 wks	14.35 [11.28, 17.41]	14.66 [10.76, 18.56]		
6 wks	13.04 [9.46, 16.62]	12.72 [7.82, 17.62]		
8 wks	11.51 [8.69, 14.33]	16.93 [13.56, 20.30]		

https://doi.org/10.1371/journal.pone.0173869.t003

Intervention Group	OutcomeMeasure	Mean Baseline Score (SD)	Mean 8-wk Score (SD)	Mean Change Score (SD)	P-value: Within-group	P-value: Between-group
Yoga (n = 15)	GSES	26.87 (3.09)	29.0 (3.89)	2.13 (2.07)	0.001	0.50
Control (n = 10)		28.5 (4.33)	30.0 (2.67)	1.5 (2.4)	0.076	
Yoga (n = 15)	RSES	14.6 (3.48)	17.47 (3.87)	2.87 (3.42)	0.006	0.053
Control (n = 10)		16.0 (3.77)	16.2 (3.88)	0.2 (3.01)	0.838	

https://doi.org/10.1371/journal.pone.0173869.t004

Figure 6.12: Above displays the lowered depressive patterns in the hatha yoga participants vs. controls (Prathikanti et al, 2017: *This is an open-access article, free of all copyright, and may be freely reproduced, distributed, transmitted, modified, built upon, or otherwise used by anyone for any lawful purpose. The work is made available under the Creative Commons CC0* public domain *dedication*).

This significant lowering of mild-to-moderate depression in adults undergoing hatha yoga becomes encouraging to employ this ancient technique to address depression: with complex etiological pathways. To quote the authors, "*This project was inspired by the need for rigorous evaluation of hatha yoga as a potential mono-therapy for individuals with mild-to-moderate major depression, a population for whom conventional care may offer uncertain benefit. With the rising popularity of yoga in western countries, yoga-based interventions with proven efficacy in major depression may provide a treatment option that is cost-effective, widely accessible, associated with high social acceptance, and has a favorable risk-benefit profile*" (Prathikanti et al, 2017).

6.7 YOGA FOR HOME-BASED ACTIVITY IN THE COVID-19 HEALTH SCARE

The quarantine imposed across the world due to the COVID-19 epidemic shall remain etched in the minds of those individuals who survived 2020! According to 2017 Centers for Disease Control and Prevention Quarantine and isolation, "*quarantine is the separation and restriction of movement of people who have potentially been exposed to a contagious disease to ascertain if they become unwell, so reducing the risk of them infecting others*".

Study ID	Country	Design	Participants	Setting	Intervention/Activities	Mental Health Outcome	Results
EXERCISE							
Legrand et al (2020) [11]	France	RCT	37 first-time prisoners with elevated anxiety	Prison	6-week interval exercise program (40 minutes/day)	Anxiety measured via State-Trait Anxiety Inventory	Exercise group (n = 20) Wait-list group (n = 17) Average reduction in state anxiety scores in the exercise group (-6.6, SD = 7.5) was greater than the average reduction in the wait-list group (-1.5, SD = 7.2) *d* = 0.69 (95%CI: .02, 1.35)
Psychou et al (2020) [12]	Greece	RCT	60 prisoners	Prison	12-week exercise program, 3 sessions per week, 60 minutes in duration	Mood states measured via Profile of Mood States (POMS) and Anxiety via The State-Trait Anxiety Inventory (STAI)	Exercise group (n = 35) Control group (n = 25) The intervention group when compared with the control group had lower mean scores on the following outcomes: Depression (1.42, SD = .63 vs 2.65, SD = .78) Total mood profile (103.24, SD = 32.3 vs 109.99, SD = 3.73) State anxiety (1.70, SD = .31 vs 2.92, SD = .53) Trait anxiety (1.96, SD = .26 vs 2.65, SD = .42) *d* Depression = 1.77 (95%CI 1.15, 2.37) State anxiety = 2.94 (95%CI: 2.19, 3.67) Trait anxiety = 2.06 (95%CI: 1.42, 2.68)
O'Toole et al (2018) [13]	Ireland	Pre-post	30 prisoners pre-existing mental health symptoms	Prison	3-month exercise program	Depression, anxiety and stress (DASS-42), Novaco Anger Scale, Rosenberg Self-Esteem Scale, Zung Self-Rated Anxiety Scale	N = 30 Post-intervention, there was a decrease in the proportion of participants with moderate to extreme scores in the following outcomes: Depression: (53.3% vs 10%) *d* = 0.99 (95%CI: 0.22, 1.75) Anxiety: (63.3% vs 20%) *d* = 0.91 (95%CI: 0.15, 1.66)
Zar et al (2017) [14]	Iran	Cross-sectional	97 prisoners	Prison	At least 3 exercise sessions each week	Anxiety, social Dysfunction, Depression measured via Goldberg & Hild's General Health Questionnaire	Exercise group (n = 44) Control group (n = 53) The intervention group when compared with the control group had lower mean scores on the following outcomes: Somatization (8.02, SD = 4.64 vs 11.16, SD = 3.52) Anxiety (7.90, SD = 4.60 vs 11.07, SD = 3.45) Social dysfunction (7.32, SD = 4.25 vs 8.66, SD = 5.43) Severe depression (6.55, SD = 5.15 vs 10.96, SD = 6.10) *d*: Somatization = 0.61 (95%CI: 0.20, 1.02) Anxiety = 0.62 (95%CI: 0.21, 1.03) Social dysfunction = 0.27 (95%CI: -0.13, 0.67) Severe depression = 0.78 (95%CI: 0.36, 1.19)
Baidawi et al (2016) [15]	Australia	Cross-sectional	173 prisoners, predominantly first-time inmates	Prison	Exercise over 4 weeks	Psychological distress measured via Kessler Psychological Distress Scale—K10	Participants who engaged in exercise in the previous month compared with those who did not exercise had, on average, 0.89 (95% CI: .25, 1.53) lower distress scores, after adjusting for potential confounders. *d* = 0.16 (95%CI: -0.19, 0.51)
Battaglia et al (2015) [16]	Italy	RCT	64 prisoners	Prison	9 months of cardiovascular plus resistance training (CRT), or high-intensity strength training (HIST)	Mental disorder symptoms using the Symptom Checklist-90-Revised (SCL-90-R)	Exercise group CRT (n = 22) Exercise group HIST (n = 22) Control group CRT (n = 20) The CRT intervention group when compared with the control group had lower mean scores on: Somatization (0.20, SD = 0.27 vs 0.23, SD = 0.24) Obsessive-compulsive (0.36, SD = 0.59 vs 0.27, SD = 0.25) Interpersonal Sensitivity (0.24, SD = 0.66 vs 0.22, SD = 0.35) Depression (0.26, SD = 0.40 vs 0.77, SD = 0.87) Anxiety (0.22, SD = 0.40 vs 0.28, SD = 0.50) Phobic anxiety (0.13, SD = 0.27 vs 0.11, SD = 0.19) Psychoticism (0.31, SD = 0.53 vs 0.24, SD = 0.29) Paranoid ideation (0.28, SD = 0.51 vs 0.43, SD = 0.46) GSI (0.26, SD = 0.40 vs 0.34, SD = 0.21) *d* (CRT): Somatization = 0.12 (95%CI: -0.49, 0.72) Obsessive-compulsive = 0.20 (95%CI: -0.41, 0.80) Interpersonal Sensitivity = 0.04 (95%CI -0.57, 0.64) Depression = 0.77 (95%CI: 0.13, 1.39) Anxiety = 0.13 (95%CI: -0.47, 0.74) Phobic anxiety = 0.08 (95%CI: -0.52, 0.69) Psychoticism = 0.16 (95%CI -0.45, 0.77 Paranoid ideation = 0.31 (95%CI: -0.30, 0.92) GSI = 0.25 (95%CI: -0.36, 0.85) The HIST intervention group when compared with the control group had lower mean scores on: Somatization (0.34, SD = 0.43 vs 0.23, SD = 0.24) Obsessive-compulsive (0.49, SD = 0.63 vs 0.27, SD = 0.25) Interpersonal Sensitivity (0.23, SD = 0.40 vs 0.22, SD = 0.35) Depression (0.24, SD = 0.27 vs 0.77, SD = 0.87) Anxiety (0.25, SD = 0.29 vs 0.28, SD = 0.50) Phobic anxiety (0.03 SD = 0.08 vs 0.11, SD = 0.19) Psychoticism (0.30, SD = 0.28 vs 0.24, SD = 0.29) Paranoid ideation (0.43, SD = 0.51 vs 0.43, SD = 0.46 GSI (0.32, SD = 0.33 vs 0.34, SD = 0.21) *d* (HIST): Somatization = 0.31 (95%CI: -0.30, 0.92) Obsessive-compulsive = 0.45 (95%CI: -0.17, 1.06) Interpersonal Sensitivity = 0.03 (95%CI: -0.58, 0.63) Depression = 0.84 (95%CI: 0.20, 1.47) Anxiety = 0.07 (95%CI: -0.53, 0.68) Phobic anxiety = 0.56 (95%CI: -0.06, 1.17) Psychoticism = 0.21 (95%CI: -0.40, 0.82) Paranoid ideation = 0.00 (95%CI: -0.61, 0.61) GSI = 0.07 (95%CI: -0.53, 0.68)
Buckaloo et al (2009) [17]	US	Cross-sectional	60 male prisoners	Prison	Exercise	Depression via Beck Depression Inventory and stress via Daily Hassles Scale	Exercise group (n = 30) No exercise group (n = 30) Mean depression scores in the exercise group (12.9, SD = 12.2) was lower compared to the no exercise group (22.5, SD = 8.8). *d* = 0.90 (95%CI = .37, 1.43) Mean scores on the DHS in the exercise group (52.6, SD = 49.0) was lower compared to the no exercise group (78.5, SD = 46.8). *d* = .56 (95%CI: 0.06, 1.17)
Cashin et al (2008) [18]	Australia	Cross-sectional	838 prisoners	Prison	Exercise over 4 weeks	Psychological well-being as measured by Beck Hopelessness Scale	Exercise was correlated with lower scores on BHS ($r = -0.079$, $p < 0.02$). *d* = .16 (95%CI: 0.02, 0.29)

YOGA (5 studies)							
Sfendla et al (2018) [19]	Sweden	RCT	152 prisoners	Prison	10-week yoga. 90-min each week	Psychological distress measured via Brief Symptom Inventory (BSI)	Yoga group (n = 77) Control group (n = 75) The intervention group when compared with the control group had lower mean scores on: GSI (0.6, SD = 0.6 vs 0.8, SD = 0.7) Anxiety (0.8, SD = 0.8 vs 0.9, SD = 0.9) Depression (0.7, SD = 0.7 vs 1.0, SD = 1.0) Interpersonal Sensitivity (0.4, SD = 0.6 vs 0.8, SD = 1.0) Obsessive-compulsive (0.8, SD = 0.7 vs 1.2, SD = 0.9) Psychoticism (0.5, SD = 0.7 vs 0.7, SD = 0.8) Paranoid ideation (0.7, SD = 0.7 vs 1.0, SD = 0.9) Phobic anxiety (0.3, SD = 0.7 vs 0.5, SD = 0.9) Somatization (0.4, SD = 0.6 vs 0.7, SD = 0.6) *d*: GSI = 0.31 (95%CI: -0.01, 0.63) Anxiety = 0.12 (95%CI: -0.20, 0.44) Depression = 0.35 (95%CI: 0.03, 0.67) Interpersonal Sensitivity = 0.49 (95%CI: 0.15, 0.81) Obsessive-compulsive = 0.50 (95%CI: 0.17, 0.81) Psychoticism = 0.27 (95%CI: -0.05, 0.59) Paranoid ideation = 0.37 (95%CI: 0.05, 0.69) Phobic anxiety = 0.25 (95%CI: -0.07, 0.57) Somatization = 0.50 (95%CI: 0.18, 0.82)
Kerekes et al (2017) [20]	Sweden	RCT	152 prisoners	Prison	10-week yoga. 90-min each week	Perceived stress measured via PSS and psychological distress measured via BSI	Yoga group (n = 77) Wait-list group (n = 75) The intervention group when compared with the wait-list group had lower mean scores on the following outcomes: Perceived stress (22.4, SD = 9.0 vs 25.7, SD = 9.3) Psychological distress (0.6, SD = 0.6 vs 0.8, SD = 0.7) *d*: Perceived stress = 0.35 (95%CI: 0.03, 0.67) Psychological distress = 0.31 (95%CI: -0.01, 0.63)
Sistig et al (2015) [21]	New Zealand	Pre-Post	26 forensic patients	Forensic Psychiatry Centre	8-week mindful yoga sessions, weekly, 60-min in duration	Anxiety and Depression measured via Hospital Anxiety and Depression Scale (HADS), PSS, The Clinical Outcomes in Routine Evaluation–Outcome Measure (CORE-OM)	Anxiety scores decreased from 7.92 (SD = 4.39), at baseline, to 7.66 (SD = 4.03), post-intervention, and to 6.67 (SD = 3.83), 2 months post-intervention. *d* = 0.11 (95%CI: -0.66, 0.87) *d* = 0.58 (95%CI: -0.21, 1.36) Depression scores increased slightly from 5.12 (SD = 3.00) to 5.23 (SD = 3.92), post-intervention, but decreased to 4.58 (SD = 3.35), 2 months post-intervention. *d* = -0.07 (95%CI: -0.84, 0.70) *d* = 0.31 (95%CI: -0.47, 1.08) Scores on perceived stress decreased slightly from 16.80 (SD = 7.98) to 16.46 (SD = 7.89), but increased to 17.51 (SD = 7.13), 2 months post-intervention. *d* = 0.09 (95%CI: -0.68, 0.86) *d* = -0.22 (95%CI: -0.98, 0.56)
Bilderbeck et al (2013) [22]	United Kingdom	RCT	100 prisoners	Prison	10-week yoga, weekly, 2 hours in duration	Perceived Stress measured via PSS, psychological distress measured via BSI, and positive and negative state affect measured via PANAS	Yoga group (n = 45) Control group (n = 55) The intervention group when compared with the control group had lower mean scores on the following outcomes: Positive affect (37.16, SD = 1.16 vs 31.22, SD = 1.02) Negative affect (15.02, SD = 0.80 vs 19.15, SD = 1.10) Perceived stress (11.40, SD = 1.10 vs 16.07, SD = 1.05) Psychological distress (24.49, SD = 3.38 vs 37.09, SD = 3.97) *d*: Positive affect = 5.47 (95%CI: 4.61, 6.33) Negative affect = 4.23 (95%CI: 3.51, 4.93) Perceived stress = 4.35 (95%CI: 3.62, 5.07) Psychological distress = 3.39 (95%CI: 2.77, 4.00)
Harner et al (2010) [23]	US	Pre-post	21 women prisoners	Prison	12-week yoga, 2 times per week, duration of 120 min	Depression, anxiety and stress as measured by Beck Depression Inventory, Beck Anxiety Inventory, Perceived Stress Scale (PSS)	At 12 weeks, depression scores decreased to 5.67 (SD = 7.5) from 21.90 (SD = 13.3) at baseline. *d* = 1.79 (95%CI: 1.07, 2.51) At 12 weeks, anxiety scores decreased to 7.3 (SD = 6.3) from 12.0 (SD = 13.8) at baseline. *d* = 0.44 (95%CI: -0.18, 1.05) Perceived stress scores were similar at 12 weeks, 22.0 (SD = 6.2) versus 22.8 (SD = 3.5) at baseline, *d* = 0.15 (95%CI: -0.46, 0.76)
OTHERS (3 studies)							
Liu et al (2020) [24]	China	RCT	51 COVID-19 patients	Hospital	5-day Progressive Muscle Relaxation (30 minutes/day)	Anxiety measured via State-Trait Anxiety Inventory	Intervention group (n = 25) Control group (n = 26) Mean state anxiety scores in the intervention group was lower (44.96, SD = 12.68) than the control group (57.15, SD = 9.24) *d* = 1.10 (95%CI: 0.51, 1.69)
Bensimon et al (2015) [25]	Israel	RCT	48 prisoners	Prison	3 weeks of listening to relaxing music, 45 minutes in duration, 3 times a day	Anxiety measured via State-Trait Anxiety Inventory (STAI), State-Trait Anger Expression Inventory (STAXI)	Music Group (n = 24) Control Group (n = 24) At week 3 of the intervention, the music group reported greater reduction (from baseline) in state anxiety than the control group (.27, 95%CI: 0.12, 0.42). *d* = 3.53 (95%CI: 2.61, 4.43) 1 week post-intervention, the greater reduction (from baseline) in state anxiety in the music group was sustained (.11, 95% CI: -0.04, 0.26). *d* = 1.44 (95%CI: 0.79, 2.07)
Cherry et al (2015) [26]	US	Cross-sectional	219 coastal residents	Community; post-disaster	Non-organizational religiosity	PTSD (17-item PTSD checklist); Depression (PHQ-9); and anxiety (GAD-7)	Increased odds of experiencing PTSD were reported among persons who were categorized as being medium, 2.03 (95% CI: 0.37, 11.3), and high, 9.8 (95%CI: 1.78, 53.78), in non-organizational religiosity when compared with persons who are low on non-organizational. d_{medium} = -0.39 (95%CI: -1.34, 0.55) d_{high} = -1.26 (95%CI: -2.20, -0.32)

*Abbreviations: BAI = Beck Anxiety Inventory; BDI = Beck Depression Inventory; BHS = Beck Hopelessness Scale; BSI = Brief Symptom Inventory; BSI-11 = Barratt Impulsiveness Scale; CORE-OM = Clinical Outcomes in Routine Evaluation- Outcome Measure; d = Cohen's d or effect size; DASS-42 = Depression Anxiety Stress Scale; GAD-7 = General Anxiety Disorder-7; GHQ = General Health Questionnaire; HADS = Hospital Anxiety and Depression Scale; K10 = Kessler Psychological Distress Scale; NAS = Novaco Anger Scale; PANAS = Positive and Negative Affect Schedule; PCL = Posttraumatic Stress Disorder Checklist; PHQ-9 = Patient Health Questionnaire-9; POMS = Profile of Mood States; PSS = Perceived Stress Scale; RSES = Rosenberg Self-Esteem Scale; SAS = Zung Self-Rated Anxiety Scale; SCL-90-R = Symptom Checklist-90-Revised; STAI = State Trait Anxiety Inventory; STAXI = State-Trait Anger Expression Inventory.

https://doi.org/10.1371/journal.pone.0243125.t001

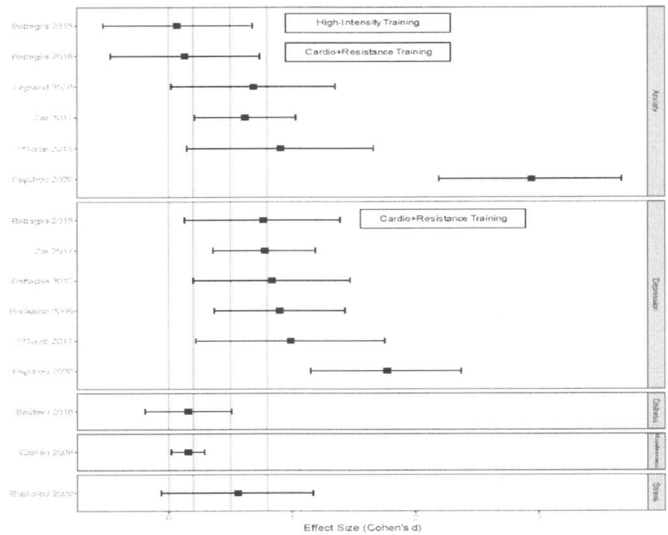

Figure 6.13: Above shows ***TOP***: studies assessing the effects of various home-activities to address the stress of the COVID-19 quarantine; ***BOTTOM***: estimating the effects of yoga (Puyat et al, 2020: *This is an open-access article distributed under the terms of the Creative Commons Attribution License, which permits unrestricted use, distribution, and reproduction in any medium, provided the original author and source are credited*).

A team of scientists reported in *The Lancet (2020)* that the psychological impacts of the quarantine ranged from "post-traumatic stress symptoms, confusion, and anger to stress about longer quarantine duration, infection fears, frustration, boredom, inadequate supplies, inadequate information, financial loss, and stigma" (Brooks et al, 2020).

Researchers Puyat and team (2020) assessed the effects of home-based activities in addressing these negative effects of the quarantine, isolation, and social distancing necessitated by the COVID-19 pandemic. Among the various techniques followed such as exercise, muscle relaxation, music, religion, and yoga; in the case of yoga, of 5 studies assessing the impact of yoga on mental health, analyses revealed that "*All point-estimates, except for one study, were above zero suggesting beneficial effects, especially for alleviating depression symptoms, distress, and stress*".

6.8 DEPRESSION AND PSYCHOSIS: THE PROMISE OF YOGA

According to 2006-published research in *World psychiatry: official journal of the World Psychiatric Association (WPA)*, historically schizophrenia and depression were considered separate entities but depression has been shown to be prevalent in 7 to 70% of psychotic patients. Further, the acute psychotic phase is associated with conspicuously depressive symptoms that can be addressed by antipsychotic interventions.

Given that the efficacy of anti-depressants and cognitive behavioral therapy for psychotic symptoms (CBTp) is still being assessed in detail, approaches termed as non-verbal have also been examined. These are inclusive of exercise, music, and yoga.

The effect of these approaches on depression in patients with psychosis was assessed in a meta-analysis conducted by researchers Steenhuis and team (2015). Analyses of randomized clinical trials of the effects of non-verbal interventions sourced from PubMed, Psychinfo, Picarta, Embase, and ISI Web of Science, up to January 2015 was done. The studies included involved 45-60 minute yoga sessions (2-8 weeks) based on asanas, pranayama, and nidra (relaxation). When compared against Treatment as usual (TAU), the effect of yoga was found to be "a significant and strong effect ($p<0.0001$)" with the ES value as -0.79 (95% CI = -1.24, -0.35) favoring yoga over TAU.

Reference	Methods	Participants*	Interventions	Outcomes†	CTAM score
Duraiswamy et al. (2007) [43]	Allocation: randomized, Blindness: single–assessors blinded, Duration: 4 months, Design: parallel group	Diagnosis: schizophrenia (DSM-IV), History: not reported, N = 61, Age: mean 32, range 18–55, Sex: 42 M, 19 F, Setting: inpatients and outpatients,	1. Yoga Therapy (breathing, relaxation and body posture exercises), 1 hour a day, 5 days a week for 3 weeks. N = 31. 2. Physical Therapy (brisk walking, jogging, postures and relaxation).1 hour a day, 5 days a week for 3 weeks. N = 30	Psychopathology (PANSS), Social and Occupational Functioning (SOFS), Side effects (Simpson Angus Scale for Extrapyramidal Symptoms & AIMS), Quality of Life (WHOQOL-BREF) Measurements: T0 and T1 (4 months),	67

Li et al. (2007) [22]	Allocation: randomized, Blindness: unknown, Duration: 6 weeks, Design: parallel group.	Diagnosis: schizophrenia (CCMD-3), History: not reported, N = 60. Age: mean 32, SD 12. Sex: 60 M, 0 F, Setting: inpatients	1. Receptive group music therapy (music listening, music listening in combination with verbal inductions given by the therapist, 'positive hypnosis'/ positive imagery,), 5 weekly 40-min sessions per week (total 30 sessions). N = 30. 2. Standard care (supportive counseling). N = 30.	Depression (SDS), Anxiety (SAS), Social functioning (NOSIE subscale), Measurements: T0 and T1 (6 weeks)	37
Lu et al. (2013) [41]	Allocation: randomized, Blindness: single—assessors blinded, Duration: 5 weeks, Design: parallel group.	Diagnosis: Schizophrenia (DSM-IV), History: not reported., N = 80, Age: mean: 52, Sex: 59 M, 21 F, Setting: inpatients.	1.) Music Therapy (active and passive music interventions, incl. music listening, singing, playing instruments, watching music videos and discussions). 60 min. sessions of therapy, twice a week, for 5 weeks (total 10 sessions). N = 38.2.) Treatment-as-usual (24 hour care, activities of daily living, nursing care, meal provision and social activities). N = 42.	Psychopathology (PANSS), Depression (CDSS), Measurements: T0, T1 (5 weeks) and T2 (3 months)	66
Manjunath et al. (2013) [44]	Allocation: randomized, Blindness: single—assessors blinded, Duration: 6 weeks, Design: parallel group.	Diagnosis: Schizophrenia (DSM-IV & MINI)., History: not reported., N = 88, Age: mean 31., Sex: 49 M, 39 F., Setting: inpatients	1.) Yoga therapy (breathing, relaxation, body posture), 1 hour daily sessions, first two weeks supervised, there-after instructed to practice on their own. N = 44 2.) Exercise therapy (general exercises), 1 hour daily sessions, first two weeks supervised, there-after instructed to practice on their own. N = 44.	Psychopathology (PANSS), Depression (HDRS), Severity of symptoms (CGIS), Extrapyramidal side-effects (SAS), Measurements: T0 and T1 (6 weeks)	42

Marzolini et al. (2009) [42]	Allocation: randomized, Blindness: single—assessors blinded, Duration: 12 weeks, Design: parallel group	Diagnosis: Schizophrenia/ schizo-affective disorder (DSM-IV & MINI), History: not reported, N = 13, Age: mean 45, Sex: 8 M, 5 F, Setting: outpatients	1.) Exercise therapy (warm up, cardiovascular exercise sessions, resistance training). 2 times a week for 90 minutes, 12 weeks total. N = 72.) Usual care (medication). N = 6	Six minute walking test (6MWD), One repetition maximum test (1RM), Anthropometric measurements (height, body mass, resting blood pressure, and waist/ hip circumference.), Adherence (attendance), Mental health inventory (MHI; subscales for anxiety, depression, behavioral control, positive affect and total score), Feedback questionnaire, Measurements: T0 and T1 (12 weeks)	38
Morgan et al. (2011) [39]	Allocation: first randomized, then quasi-randomized, Blindness: single—assessors blinded, Duration: 2 weeks, Design: parallel group	Diagnosis: schizophrenia, schizoaffective disorder or bipolar affective disorder (DSM-IV), History: not reported, N = 49, Age: mean: 36, range 17–55, Sex: 23 M, 26 F, Setting: outpatients,	1.) Music therapy (improvisation or song writing), 4 individual sessions over two weeks. Between 10–30 minutes. N = 252.) Active control (listen to a CD with relaxing sounds), 4 individual sessions over two weeks. Between 10–30 minutes. N = 24	Brief Psychiatric Rating Scale (BPRS), Calgary Interview Guide for Depression, Nurses Observation Scale for Inpatient Evaluation (NOSIE-30), Depression Anxiety Stress Scale (DASS 21), Measurements: T0, T1 (2 weeks) and T2 (6 weeks).	46
Scheewe et al. (2013) [45]	Allocation: randomized, Blindness: single—assessors blinded, Duration: 6 months, Design: parallel group	Diagnosis: schizophrenia, schizoaffective disorder or schizofreniform disorder (DSM-IV; confirmed with CASH), History: not reported, N = 63, Age: mean: 29, Sex: 46 M, 17 F, Setting: outpatients	1.) Exercise therapy (Cardiovascular exercises & muscle strength exercises), an hour of exercise twice a week for six months. N = 31.2.) Occupational therapy (creative and recreational activities, like painting or reading), N = 32.	Psychopathology (PANSS), Depression (MADRS), Need of care (CAN), Cardiorespiratory fitness (CPET), BMI, Body fat percentage (BFP), Measurements: T0 and T1 (6 months)	78
Ulrich et al. (2007) [38]	Allocation: randomized, Blindness: single—assessors blinded;assessors unaware of study aim; success of blindingverified, Duration: 4.8 weeks, Design: parallel group.	Diagnosis: schizophrenia or related psychoses (27 of 37 had F20 in ICD-10), History: not reported, N = 37, Age: mean 38, range 22–58, Sex: 20 M, 17 F, Setting: inpatients,	1. Active group music therapy (focusing on musical processes and discussion of patients'problems), on average 7.5 sessions of 60 to 105 minutes. N = 21.2. Standard care (medication, "other" activities—no detailed description given). N = 16	Mental state: SANS, Quality of life: SPG, Unable to use, Social functioning (unvalidated subscale of published scale), Satisfaction with care (unpublished scale), Measurements: T0 and T1 (5 weeks).	58

Visceglia & Lewis (2011) [40]	Allocation: randomized, Blindness: single—assessor blinded, Duration: 8 weeks, Design: parallel group.	Diagnosis: schizophrenia (DSM-IV). However, many participants had multiple diagnoses, History: not reported, N = 18, Age: mean 42 years, Sex: 12 M, 6 F, Setting: inpatients	1. Yoga therapy (breathing exercises, warm-ups, postures, and deep relaxation), 45 min twice weekly sessions for 8 weeks (total of 16 sessions). N = 10.2. Waitlist (medication—no detailed description given). N = 8	Psychopathology (PANSS), Quality of life (WHOQOL-BREF), Measurements: T0 and T1 (8 weeks)	40
Wen et al. (2005) [23]	Allocation: randomized, Blindness: unknown, Duration:6 weeks, Design: parallel group.	Diagnosis: schizophrenia (CCMD-3), History: not reported, N = 30, Age:15 to 50, Sex: 21 M, 9 F, Setting: inpatients	1. Receptive group music therapy (music listening, other music activities: dancing, discussion emphasizing the emotional aspects of the music while listening to it), five one hour sessions per week (total 30 sessions). N = 16.2. Standard care (medication only, no anxiolytic or antidepressant). N = 14	Mental state: BPRS; depression (SDS, HDRS), Unable to use—Inpatient Recovery Effect Scale (unpublished scale), Measurements: T0 and T1 (6 weeks).	32

Figure 6.14: Above shows ***TOP***: the 10 studies utilized in the meta-analyses of the effect of non-verbal therapies on depression in psychosis; ***BOTTOM***: Network comparisons of the non-verbal approaches (Steenhuis et al, 2015: *This is an open-access article distributed under the terms of the Creative Commons Attribution License, which permits unrestricted use, distribution, and reproduction in any medium, provided the original author and source are credited*).

Overall, it was concluded that non-verbal therapies diminish the depressive aspects of psychotic diseases-these are inclusive of yoga and music therapy opening up the avenue of exploring these approaches to address this serious disease.

(Steenhuis et al, 2015)

6.9 FINAL WORDS

The idea that yoga is only n exercise is being replaced by the idea that its use as complementary and alternative medicine is safe and reliable. While its utility in healing and addressing several diseases is being unveiled by research, this chapter outlined the efficacy of yoga to alleviate depressive symptoms. According to Sivaramakrishnan et al (2019) who reported the benefits of yoga in elderly people, "*This review adds to the growing evidence on the benefits of yoga, and researchers should work closely with yoga teachers, studios, fitness centres and policy makers to develop and implement strategies to encourage yoga participation among older adults, tying in with the final aim of increasing participation in muscle strength and balance activities*".

Addressing major depression in select populations such as the elderly, adolescents, and sensitive people using yoga as a non-pharmacological therapeutic option looks exciting. Further studies in this direction are warranted. Further analyses of how yoga fares as a mono-therapy to address depression against antidepressants can be another option. Janakiramaiah et al (2000) reported that the total scores on Beck Depression Inventory (BDI) and Hamilton Rating Scale for Depression (HRSD) were on the same lines for 45 people with depression on imipramine and Sudarshan Kriya Yoga (SKY).

There is also scope for developing tailor-made yoga practices for depression based on the symptoms expressed by an individual. This can entail vigorous yoga for those with hopelessness and emotional numbing and yoga relaxation for those with insomnia or restlessness. Given that nascent research can predict whether an individual will respond to psychotherapy or anti-depressants to thus bring in adopting yoga for those who may not respond to conventional approaches.

Given the host of factors involved in the etiology of depression, the use of yoga is not aimed to completely replace the current therapeutic approaches but to broaden the scheme of treatment to allow more individuals to be healed!

(Schreiber and Niemeier, 2013; Prathikanti et al, 2017; Sivaramakrishnan et al, 2019)

6.10 REFERENCES

1. Shohani, M., Badfar, G., Nasirkandy, M. P., Kaikhavani, S., Rahmati, S., Modmeli, Y., Soleymani, A., & Azami, M. (2018). The Effect of Yoga on Stress, Anxiety, and Depression in Women. *International journal of preventive medicine*, *9*, 21. https://doi.org/10.4103/ijpvm.IJPVM_242_16
2. Theresa A. Schreiber, Brandi S. Niemeier (2013). Department of Health, Physical Education, Recreation & Coaching, University of Wisconsin-Whitewater, 130 Williams Center, 800 West Main Street, Whitewater, WI 53190. Exploring the Health Benefits of Yoga: A Review. World Journal of Medical Research Volume No 8.
3. Erik J Groessl, Deepak Chopra and Paul J Mills. An Overview of Yoga Research for Health and Well-Being. J Yoga Phys Ther 2015, 5:4 DOI: 10.4172/2157-7595.1000210.
4. Woolery A, Myers H, Sternlieb B, Zeltzer L. A yoga intervention for young adults with elevated symptoms of depression. Altern Ther Health Med. 2004 Mar-Apr;10(2):60-3. PMID: 15055096.
5. Smith JA, Greer T, Sheets T, Watson S. Is there more to yoga than exercise? Altern Ther Health Med. 2011 May-Jun;17(3):22-9. PMID: 22164809.
6. Stussman BJ, Black LI, Barnes PM, Clarke TC, Nahin RL. Wellness-related Use of Common Complementary Health Approaches Among Adults: United States, 2012. Natl Health Stat Report. 2015 Nov 4;(85):1-12. PMID: 26556396.
7. Chaturvedi S, Chandra PS, Channabasavanna SM, Anantha N, Reddy BK, Sharma S. Levels of anxiety and depression in patients receiving radiotherapy in India. Psychooncology. 1996;5:343–6.
8. Chandwani, K. D., Perkins, G., Nagendra, H. R., Raghuram, N. V., Spelman, A., Nagarathna, R., Johnson, K., Fortier, A., Arun, B., Wei, Q., Kirschbaum, C., Haddad, R., Morris, G. S., Scheetz, J., Chaoul, A., & Cohen, L. (2014). Randomized, controlled trial of yoga in women with breast cancer undergoing radiotherapy. *Journal of clinical oncology : official journal of the American Society of Clinical Oncology*, *32*(10), 1058–1065. https://doi.org/10.1200/JCO.2012.48.2752.
9. Rao, R. M., Raghuram, N., Nagendra, H. R., Usharani, M. R., Gopinath, K. S., Diwakar, R. B., Patil, S., Bilimagga, R. S., & Rao, N. (2015). Effects of an integrated Yoga Program on Self-reported

Depression Scores in Breast Cancer Patients Undergoing Conventional Treatment: A Randomized Controlled Trial. *Indian journal of palliative care*, *21*(2), 174–181. https://doi.org/10.4103/0973-1075.156486

10. M. Horneber, G. Bueschel, G. Dennert, D. Less, E. Ritter, and M. Zwahlen, "How many cancer patients use complementary and alternative medicine: a systematic review and metaanalysis," *Integrative Cancer Therapies*, vol. 11, no. 3, pp. 187–203, 2012.
11. Marcy McCall, Melanie McDonald, Sally Thorne, Alison Ward, Carl Heneghan, "Yoga for Health-Related Quality of Life in Adult Cancer: A Randomized Controlled Feasibility Study", Evidence-Based Complementary and Alternative Medicine, vol. 2015, Article ID 816820, 12 pages, 2015. https://doi.org/10.1155/2015/816820
12. Michael J. Mackenzie, Amanda J. Wurz, Yayoi Yamauchi, Lanie Ann Pires, S. Nicole Culos-Reed, "Yoga Helps Put the Pieces Back Together: A Qualitative Exploration of a Community-Based Yoga Program for Cancer Survivors", Evidence-Based Complementary and Alternative Medicine, vol. 2016, Article ID 1832515, 10 pages, 2016. https://doi.org/10.1155/2016/1832515
13. Sivaramakrishnan, D., Fitzsimons, C., Kelly, P. *et al.* The effects of yoga compared to active and inactive controls on physical function and health related quality of life in older adults- systematic review and meta-analysis of randomised controlled trials. *Int J Behav Nutr Phys Act* **16,** 33 (2019). https://doi.org/10.1186/s12966-019-0789-2.
14. Jain V, Sharma J (2017), Effect of Yogic Intervention: Pranayama on Anxiety & Depression, International Journal of Indian Psychology, Volume 4, Issue 3, ISSN:2348-5396 (e), ISSN:2349-3429 (p), DIP:18.01.053/20170403.
15. Uebelacker LA, Epstein-Lubow G, Gaudiano BA, Tremont G, Battle CL, Miller IW. Hatha yoga for depression: critical review of the evidence for efficacy, plausible mechanisms of action, and directions for future research. J Psychiatr Pract. 2010 Jan;16(1):22-33. doi: 10.1097/01.pra.0000367775.88388.96. PMID: 20098228.
16. Balasubramaniam M, Telles S and Doraiswamy PM (2013) Yoga on our minds: a systematic review of yoga for neuropsychiatric disorders. *Front. Psychiatry* **3**:117. doi: 10.3389/fpsyt.2012.00117
17. Vollbehr NK, Bartels-Velthuis AA, Nauta MH, Castelein S, Steenhuis LA, Hoenders HJR, et al. (2018) Hatha yoga for acute, chronic and/or treatment-resistant mood and anxiety disorders: A systematic

review and meta-analysis. PLoS ONE 13(10): e0204925. https://doi.org/10.1371/journal.pone.0204925

18. Prathikanti S, Rivera R, Cochran A, Tungol JG, Fayazmanesh N, Weinmann E (2017) Treating major depression with yoga: A prospective, randomized, controlled pilot trial. PLoS ONE 12(3): e0173869. https://doi.org/10.1371/journal.pone.0173869.
19. Centers for Disease Control and Prevention Quarantine and isolation. https://www.cdc.gov/quarantine/index.html; Date: 2017
20. Brooks SK, Webster RK, Smith LE, Woodland L, Wessely S, Greenberg N, et al. The psychological impact of quarantine and how to reduce it: rapid review of the evidence. The Lancet. 2020;395: 912–920. pmid:32112714
21. Puyat JH, Ahmad H, Avina-Galindo AM, Kazanjian A, Gupta A, Ellis U, et al. (2020) A rapid review of home-based activities that can promote mental wellness during the COVID-19 pandemic. PLoS ONE 15(12): e0243125. https://doi.org/10.1371/journal.pone.0243125.
22. Oosthuizen, P., Emsley, R., Niehaus, D., Koen, L., & Chiliza, B. (2006). The relationships between depression and remission in first-episode psychosis. World psychiatry : official journal of the World Psychiatric Association (WPA), 5(3), 172–176.
23. Steenhuis LA, Nauta MH, Bocking CLH, Pijnenborg GHM (2015) Treating Depressive Symptoms in Psychosis: A Network Meta-Analysis on the Effects of Non-Verbal Therapies. PLoS ONE 10(10): e0140637. https://doi.org/10.1371/journal.pone.0140637.
24. Janakiramaiah N, Gangadhar BN, Naga Venkatesha Murthy PJ, Harish MG, Subbakrishna DK, Vedamurthachar A. Antidepressant efficacy of Sudarshan Kriya Yoga (SKY) in melancholia: a randomized comparison with electroconvulsive therapy (ECT) and imipramine. J Affect Disord. 2000 Jan-Mar;57(1-3):255-9. doi: 10.1016/s0165-0327(99)00079-8. PMID: 10708840.

CHAPTER 7

MEDITATION, MINDFULNESS AND MANTRAS AS SOLUTIONS

CONTENTS

7.1 RESEARCH AND MEDITATION

"When we refer to a journey... On the one hand we are talking about nonbeing, no world, nonexistence; and on the other hand we are discussing the process of the path, how we could proceed along a path and exert ourselves... If we split hairs in that way, there is no truth anywhere, none whatsoever. Let it just be that way; let us have contradictions"-Trungpa (2010)

The trend towards meditation with its origin in the Vedic texts of India is picking up much sought after across the globe. This is based on the concept of a deep inner self ornon-changing pure consciousness that activates the working consciousness composed of the ego, mind, intellect and "chitta" (memory bank).

"*According to Vedic* science, the deep inner Self activates the inner faculty (working consciousness), which in turn activates the physical body. A feedback loop is provided by meditation, *in which a conscious connection is made with the deep inner Self. This view of the human being correlates with the scientific view of deoxyribonucleic acid (DNA) in the body. On the cellular level, DNA creates and controls all activities in the body. Information from the DNA proceeds to ribonucleic acid (RNA), then to the amino acids, through which proteins are formed. A feedback loop to the DNA starts a new cycle to provide whatever is needed for the activities of the cell. In meditation, the feedback loop to the deep inner Self (the seat of knowledge, like DNA) provides inner peace and bliss, which removes the accumulated stresses of life and improves overall health*" (Sharma, 2015).

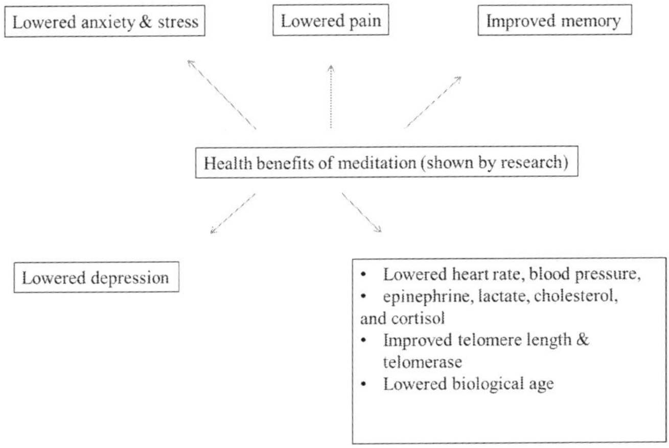

Figure 7.1: Several health benefits have been shown by meditation (Based on Sharma et al, 2015).

7.2 MINDFUL MEDITATION: THE BRAIN CHANGES

Mindfulness meditation is a form of meditation where experiences of the present moment are given full attention without any judgment. Based on Buddhist meditation practices, the entire attention is on the present moment by distancing oneself from internal and external experiences to target emotional reactions.

The efficacy of an amalgamation of cognitive behavioral training and meditation or mindfulness-based cognitive therapy (MBCT) in quelling the relapse or recurrence of major depressive disorder (MDD) was evaluated employing a meta-analysis. The 2011-published work based on Cochrane Collaboration Review Manager 5.1 analyses narrowed down 6 randomized controlled trials with a total of 593 participants. As against placebos or controls, the risk of relapse or recurrence was 34%. Interestingly, the authors reported that "In two studies, MBCT was at least as effective as maintenance antidepressant medication" (Piet and Hougaard. 2011)

The effect of mindfulness was assessed by researchers Murakami and team (2012) employingthe Five Facet Mindfulness Questionnaire (FFMQ), based on five facets on 19 student volunteers (undergraduate and graduate; mean 21.2 years in age). Non-reactivity: nonreactivity to inner experience; observing: noticing and paying attention to one's own emotions, feelings, body experience, and behavior; acting with awareness/automatic pilot/ concentration/non-distraction; describing: finding the words to describe one's own feelings; non judging of experience are the facets of the FFMQ. The study also involved estimating the regional gray matter volume in the human brain using MRI. As seen with earlier work on those who meditated, a higher score of mindfulness was linked to increased right anterior insula volume. This part of the brain is associated with emotional awareness and giving substance to emotions.

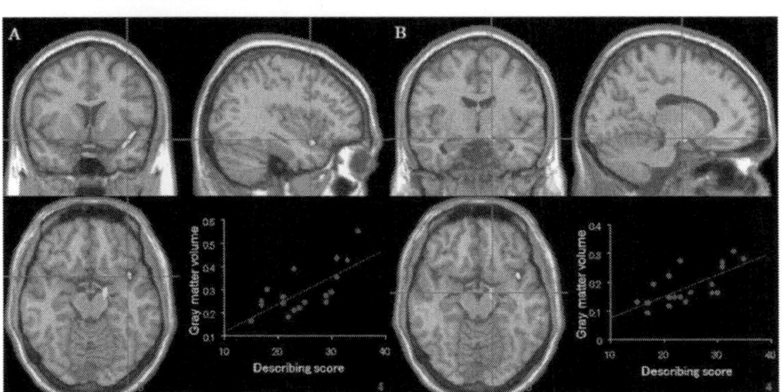

Region	Brodmann areas	Side	Coordinates			Cluster size	*t* value
			x	*y*	*z*		
Correlated with describing facet							
Anterior insular cortex	13	Right	37	10	−15	442	3.72
Parahippocampal gyrus/amygdala	28	Right	13	−7	−14	192	3.53

doi:10.1371/journal.pone.0046377.t001

Figure 7.2: Above summarizes the findings of gray matter volume and mindfulness. ***TOP***: Gray matter right anterior insula volume and mindfulness; ***BOTTOM***: Mindfulness and gray matter (Murakami et al, 2012; *This is an open-access article distributed under the terms of the Creative Commons Attribution License, which permits unrestricted use, distribution, and reproduction in any medium, provided the original author and source are credited*).

The effects of meditation were assessed in a 2013-published work described as "the first study to utilize magnetic resonance spectroscopy to evaluate the brain metabolite patterns in long-term meditators, to compare them with healthy non-meditating individuals and to elucidate the possible association between meditators' brain changes and years of meditation". The subjects were 10 meditators from the Soto Zen Spanish Buddhist community against healthy controls (non-meditators) subjected to Magnetic resonance imaging (MRI), Magnetic resonance spectroscopy (MRS), Diffusion-weighted imaging (DWI), and Diffusion tensor imaging (DTI).

The comparison of the Hospital Anxiety Depression Scale (HADS) revealed that the level of mindfulness was more while scores of depression were lower in the meditators against the non-meditators. MRS analyses revealed that Glutamate (p=.004), N-acetyl-aspartate (p=.003), and N-acetyl-aspartate/Creatine (p=.002) levels were lowered with higher myo-inositol (ml) in the posterior cingulate gyrus in the meditators. Years of meditation were found to be significantly negatively correlated with Glu (r=−0.452; p=.045), NAA (r=−0.617; p=.003) and NAA/Cr (r=−0.448; p=.047) and positively correlated with mI (myo-inositol) in the posterior cingulate cortex.

Psychological variables	Non-meditator healthy controls (N = 10)	Meditators (N = 10)	Significance
Anxiety (HADS-anx) (mean, SD)	1.70 (0.94)	0.30 (0.48)	p = 0.002
Depression (HADS-dep) (mean, SD)	1.60 (0.51)	0.30 (0.48)	p = 0.001
Cognitive function (MMSE) (mean, SD)	35 (0)	35 (0)	p = 1
Mindfulness (MAAS) (mean, SD)	19.1 (2.96)	74.1 (5.32)	p = 0.001
Months of meditation in meditators (mean, SD, range)	------	190.80 (91.81); range: 96–360 months	---

Mann-Whitney U.
doi:10.1371/journal.pone.0058476.t001

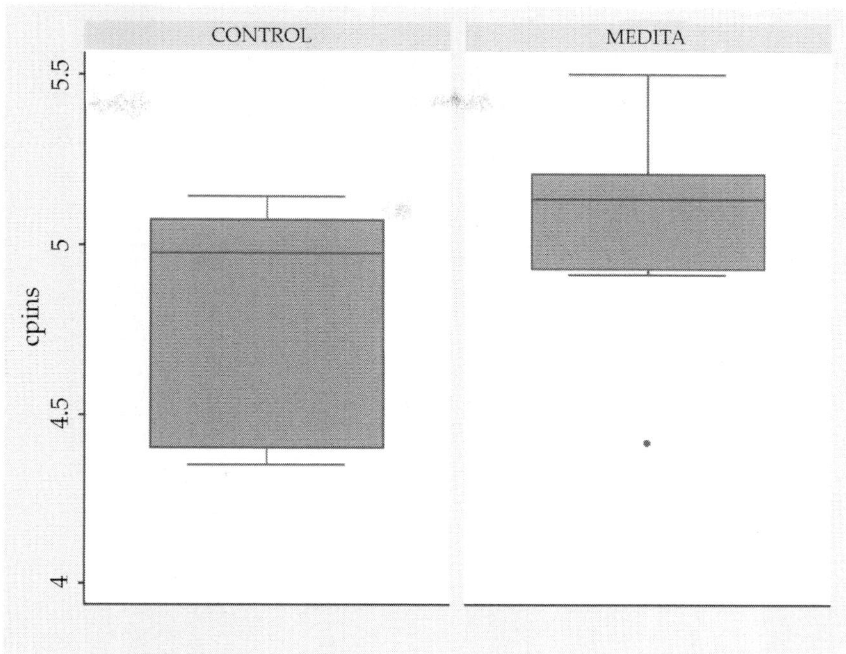
CONTROL
MEDITA
cpins
5.5
5
4.5
4

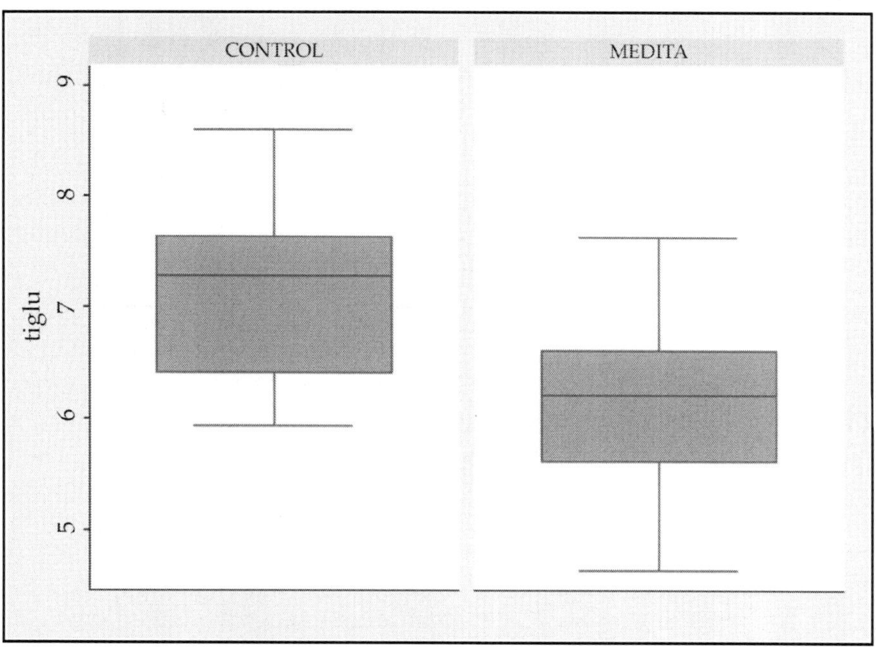
CONTROL
MEDITA
tiglu
9
8
7
6
5

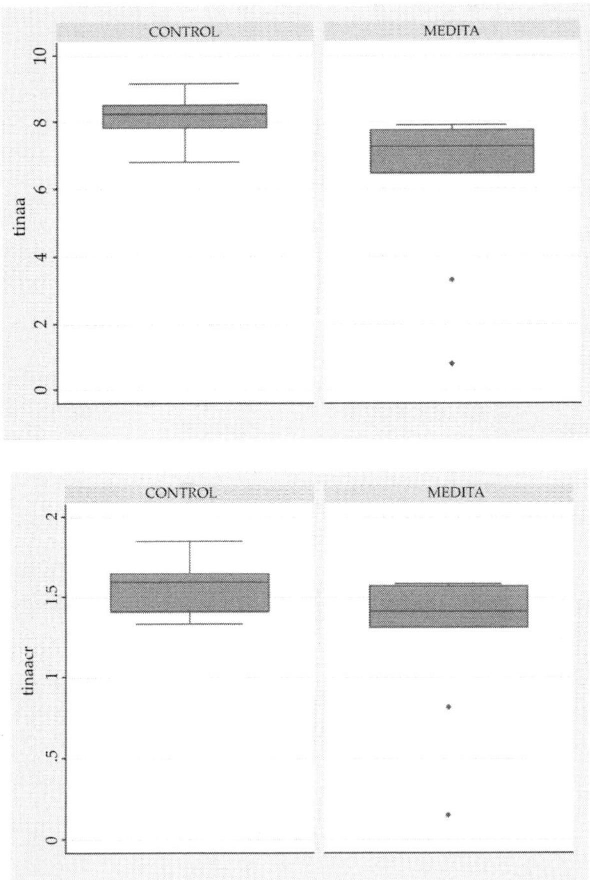

Region and Metabolites	Correlation (r)	*p*
Posterior Cingulate		
Myo-Inosotol	0.518	0.019
Left Thalamus		
Glutamate	−0.452	0.045
NAA	−0.617	0.003
NAA/Creatine	−0.448	0.047

doi:10.1371/journal.pone.0058476.t003

Figure 7.3: Above summarizes the patterns of brain metabolites in meditators: Differencesin depression scores, myo-inositol levels, glutamate levels, NAA levels, NAC/creatinine levels and, correlation assessments: (Fayed et al, 2013: *This is an open-access article distributed under the terms of the Creative Commons Attribution License, which permits unrestricted use, distribution, and reproduction in any medium, provided the original author and source are credited*).

The higher levels of myo-inositol detected in this work become significant given that unipolar and bipolar depressed patients had lowered inositol in their CSF with the administration of inositol lowering of the Hamilton Depression Rating Scale score. The meditators displayed lower mean diffusivity (MD) and increased myelination to augment communication(Fayed et al, 2013).

Continuing the analyses of how such Mindfulness Based Stress Reduction (MBSR) can quell depression, anxiety, and stress, MRI analyses with psychological evaluation on 24 participants was reported by scientists Santarnecchi and team (2014). An 8-week MBSR was administered to the test participants that was inclusive of "practicing body scanning, sitting meditation, walking meditation and mindful stretching movements" followed by psychological evaluation employing the Toronto Alexithymia Scale (TAS-20), Penn State Worry Questionnaire (PSWQ), State-Trait Anxiety Inventory (STAI) - Form Y, Beck Depression Inventory-II (BDI II), and Mindful Attention Awareness Scale (MAAS).

The level of depression was lowered: $t_{(22)}=2.086$; $p<.046$ following the mindfulness meditation as against controls. This was also accompanied by a dip in worry ($t_{(22)}=2.665$; $p<.012$), state anxiety ($t_{(22)}=2.259$; $p<.031$), and alexithymia [difficulty in emotion recognition] ($t_{(22)}=3.142$; $p<.004$). The meditation participants also demonstrated an increased cortical thickness in the right insula when compared with the controls. Those who meditated were shown to be identified by brain changes with 89% specificity by optimized voxel-based morphometry (VBM).

	MBSR				Wait-list			p
	T0 (M±SD)	T1	F	p	T0 (M±SD)	T1	F	
Age	31±4	-	-	-	30±4	-	-	-
Gender	12 F, 11 M	-	-	-	11 F, 11 M	-	-	-
Total brain volume (cm³)	1467±232	-	-	-	1398±255	-	-	-
MAAS	62±16	65±10	-.534	.597	63±6	62±9	.312	.558
PSWQ	43±10	35±8	2.665	*.012*	41±1	40±2	.347	.260
STAI_State	38±9	32±6	2.259	*.031*	37±2	36±6	.345	.220
STAI_Trait	40±9	38±8	.716	.479	39±2	39±4	.233	.370
TAS	48±14	34±12	3.142	*.004*	45±3	44±6	.413	.450
BDI	9±5	5±6	2.086	*.046*	7±4	7±2	.134	.480

Table 1 shows demographic data and pre (t0) - post (t1) MBSR changes in psychological dimensions of both groups of participants. P-values refer to a repeated measure ANCOVA ($p<.05$).
doi:10.1371/journal.pone.0108359.t001

Panels A, B, C and D below display the comparison of the voxel-wise cortical thickness between MBSR trainees and controls:

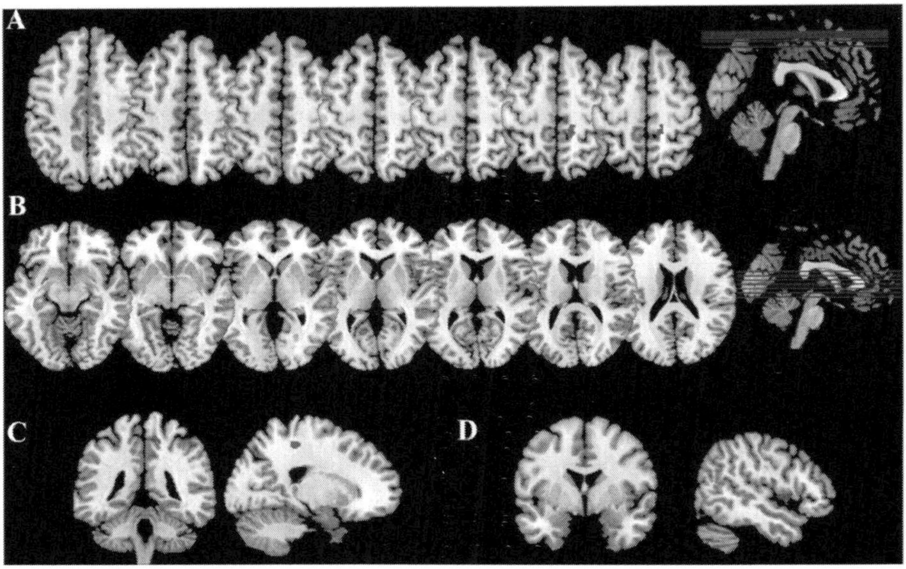

Panel A below represents the average cortical thickness values of the insula and somatosensory cortex following the mindful meditation while B shows the correlation between the cortical thickness of the insula (y axis) and alexithymia level in the meditation participants (x axis):

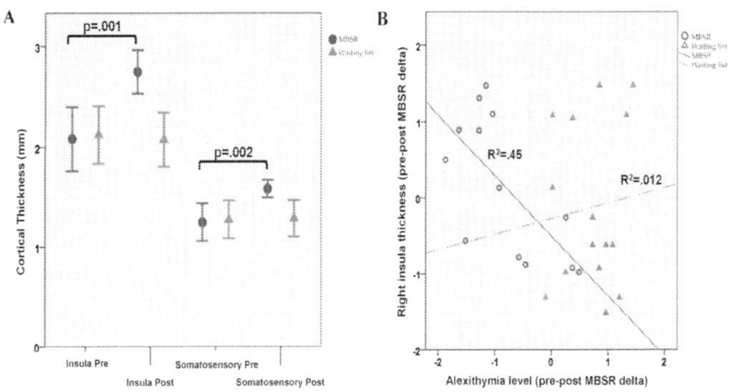

Panels A and B below show the surface-based representation of average cortical thickness in the meditating participants before and after the program using machine learning. The regions facilitating discrimination are displayed below (red line=90th° percentile); ANG=Angular gyrus; PCL=Paracentral lobule; SFGmed=Superior frontal gyrus medial part; SFGdor=Superior frontal gyrus dorsal part; ACG=Anterior cingulate gyrus; INS=Insula lobe; PHG=Parahippocampal gyrus; TPOsup=superior Temporal pole.

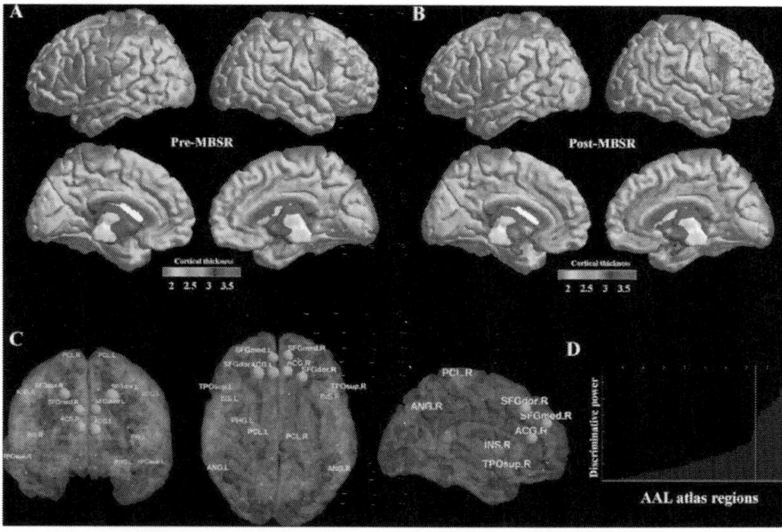

Figure 7.4: Above displays the results of an 8-week meditation-based therapy on depression. From top to bottom: Changes in the psychological profile, increased cortical thickness, and the identification (83.32% accurate) of a surface-based representation of average cortical thickness values in the meditators (Santarnecchi et al 2014: *This is an open-access article distributed under the terms of the Creative Commons Attribution License, which permits unrestricted use, distribution, and reproduction in any medium, provided the original author and source are credited*).

The researchers concluded that "*Taken together, these findings documented the significant impact of a brief MBSR training on brain structures, as well as stressing the idea of MBSR as a valuable tool for alexithymia modulation, also originally providing a plausible neurobiological evidence of a major role of right insula into mediating the observed psychological changes*".

The effects of mindfulness on worry, rumination, reappraisal, and suppression: 4 aspects of emotional healthwere assessed by Parmentier and team (2019). Worry described as "a chain of thoughts and images, negatively affect-laden, and relatively uncontrollable; it represents an attempt to engage in mental problem-solving on an issue of which outcome is uncertain but contains the possibility of one or more negative outcomes" (Borkovec et al., 1983, p. 10) is an integral part of anxiety. Depression commonly involved rumination that is thinking of past negative experiences along with their causes and effects in a repeated format. Reformulating the meaning of a situation constitutes reappraisal and suppression is inhibiting the expression of emotions but not their experience. The former helps to quell depression and anxiety while the latter approach is associated with depressive disorders.

Of 1151 adults (900 women) aged 18 to 76participating in the study entailing a voluntary online survey, meditation was reported to be practiced by 42.7%. The measures employed were Mindfulness: Short Five-Facet Mindfulness Questionnaire (FFMQ), Depression and Anxiety: Hospital Anxiety and Depression Scale (HADS), Rumination: Leuven Adaptation of the Rumination on Sadness Scale (LARSS), Worry: Penn State Worry Questionnaire (PSWQ), and Emotional Regulation: Emotion Regulation Questionnaire (ERQ).

Analyses revealed that depression and anxiety were directly and indirectly affected by mindfulness. Mindfulness lowered expressive suppression, worry, and rumination (positively correlated with depression) while increasing cognitive reappraisal (negatively correlated with depression).

Variable	***M***	***SD***	**Min**	**Max**
Estimated hours of meditation	447.739	1860.453	0	30940
Mindfulness (FFMQ)	79.511	11.475	47	112
Emotive Suppression (ERQ-S)	13.036	5.487	4	28
Cognitive Reappraisal (ERQ-R)	27.817	6.867	6	42
Rumination (LARSS)	41.637	16.296	17	85
Worry (PSWQ)	7.956	3.199	3	15
Depression (HADS-D)	3.904	3.155	0	16
Anxiety (HADS-A)	7.168	3.727	0	19

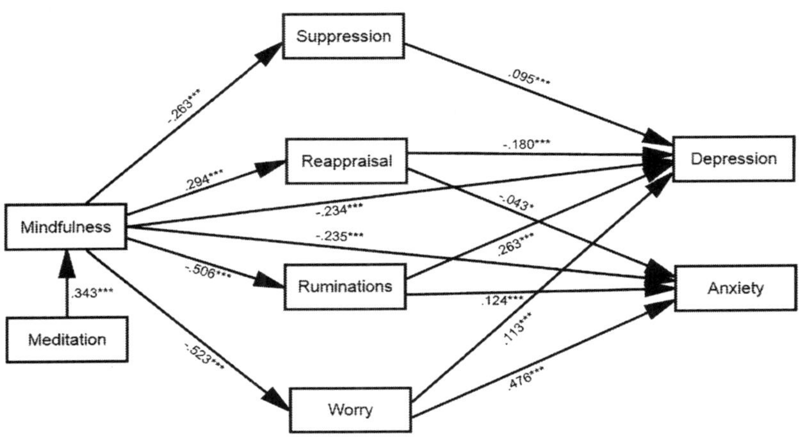

Predictor	*M*	*SE*	**Lower**	**Upper**	*p*
Effect of meditation on mindfulness					
Meditation	0.343	0.028	0.287	0.395	<0.001
Effect of mindfulness on mediators					
Reappraisal	0.294	0.030	0.232	0.352	<0.001
Suppression	−0.263	0.029	−0.321	−0.206	<0.001
Rumination	−0.506	0.024	−0.553	−0.460	<0.001
Worry	−0.523	0.023	−0.566	−0.476	<0.001
Effect of mediating and control variables on depression					
Mindfulness	−0.234	0.031	−0.296	−0.173	<0.001
Reappraisal	−0.180	0.026	−0.229	−0.128	<0.001
Suppression	0.095	0.025	0.046	0.145	<0.001
Rumination	0.263	0.032	0.199	0.323	<0.001
Worry	0.113	0.033	0.048	0.177	0.001
Age	0.100	0.026	0.050	0.152	<0.001
Gender	0.054	0.024	0.007	0.102	0.027
Language	−0.060	0.028	−0.115	−0.005	0.032
Effect of mediating and control variables on anxiety					
Mindfulness	−0.235	0.026	−0.283	−0.184	<0.001
Reappraisal	−0.043	0.023	−0.088	0.001	0.047
Rumination	0.124	0.026	0.072	0.175	<0.001
Worry	0.476	0.026	0.426	0.527	<0.001
Age	−0.057	0.023	−0.104	−0.012	0.010
Language	−0.126	0.022	−0.167	−0.083	<0.001

Figure 7.5: Above summarizes the effects of mediation-based meditation in lowering depression to augment well-being. **TOP** table: participant details (n=1151); the **chart** is representative of the association of depression with mindfulness while the **last table** shows the association of meditation and mindfulness and depression(Parmentier et al, 2019: This is an open-access article distributed under the terms of the Creative Commons Attribution License (CC BY). The use, distribution or reproduction in other forums is permitted, provided the original author(s) and the copyright owner(s) are credited and that the original publication in this journal is cited, in accordance with accepted academic practice. No use, distribution or reproduction is permitted which does not comply with these terms).

The study thus showed that meditation augments mindfulness to target and lower depression and augment overall health. A major mediator here is healthy emotional regulation. Opening up the avenue of utilizing meditation to address depression, the researchers stated "*Consequently, with respect to strategies for reducing symptoms of depression and anxiety in the general population and help reduce the risk of sub-clinical symptoms converting to clinical disorders, mindfulness-based interventions or meditation exercises focusing specifically on the reduction of worry and rumination may prove especially efficient and beneficial as potential protectors against depression and anxiety in the general population*" (Parmentier et al, 2019).

7.3 MULTIPLE SCLEROSIS, DEPRESSION, AND MEDIATION

Targeting close to 1 million people worldwide, multiple sclerosis (MS) is described as a chronic demyelinating disease of the central nervous system. The annual prevalence of depression in MS patients is 16% with a prevalence of 19 to 54%, based on the diagnosis approaches and the population. For a pediatric population, 27% display depressive signs. In a nutshell, patients with MS display depression at prevalence more than other conditions or the general public. A cross-sectional community study of MS patients revealed that 59% of the depressed patients were not on medication while this value was 65.6% in another study. As depression can be addressed, these studies point out the necessity of treating depression in MS patients to augment their life quality(Skokou et al, 2012).

Early work bySimmons and team (2004) described as "an anonymous, self-report, Internet-based survey" revealed that of 2529 people with MS, meditation accounted as a beneficial factor along with cold baths and diet. Esmonde and Long (2008) reported that attendees at a National Congress of the MS Trust in the UK labeled meditation, yoga, massage, reflexology as 'extremely helpful' in complementary and alternative medicine (CAM) to help in symptom management of MS.

Researchers Levin and team (2014) as part of the Health Outcomes and Lifestyle Interventions in a Sample of People with Multiple Sclerosis" (the HOLISM study) looked at meditation and the health-related quality of life (HRQOL) and depression in MS patients. Based on an online survey, the Multiple Sclerosis Quality of Life-54 survey (MSQOL-54) and the two-item Patient Health Questionnaire-2 (PHQ-2: a shortened version of the PHQ-9) were employed to evaluate theHRQOL and depression, respectively. The

12-month practice of meditation was also assessed on the basis of "ever, less than once per week, once to twice a week, three to four times a week, five to six times a week, and every day".

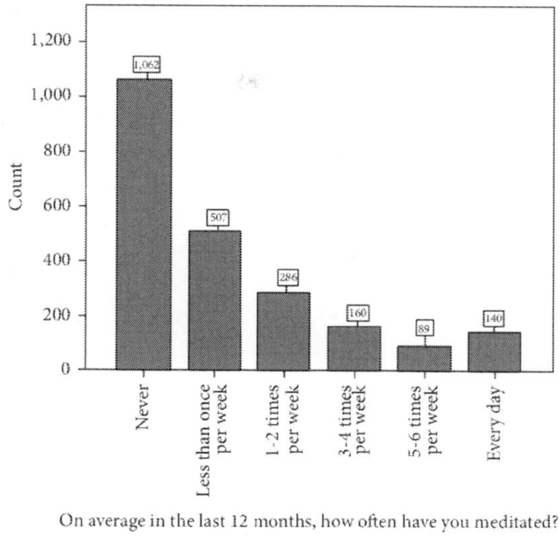

Dependent	Covariates	B	95% CI Lower	95% CI Upper	P	Adjusted R^2
MHC	Age, years	0.10	0.02	0.19	0.022	0.030
	Gender, male	0.62	−1.78	3.02	0.613	
	Secondary school or less[*]	−7.31	−9.94	−4.67	<0.001	
	Vocational training[*]	−4.45	−7.42	−1.49	0.003	
	Bachelor degree[*]	−1.64	−4.03	0.75	0.179	
	Meditates once or more per week[†]	4.80	2.65	6.94	<0.001	
	Meditates less than once per week[†]	0.78	−1.55	3.11	0.513	
PHC	Age, years	0.42	0.33	0.51	<0.001	0.081
	Gender, male	2.67	0.22	5.11	<0.001	
	Secondary school or less[*]	−8.60	−11.38	−5.82	<0.001	
	Vocational training[*]	−7.31	−10.37	−4.24	<0.001	
	Bachelor degree[*]	−2.16	−2.71	2.28	0.87	
	Meditates once or more per week[†]	2.99	0.75	5.24	0.009	
	Meditates less than once per week[†]	2.69	0.24	5.15	0.031	
		OR	Lower	Upper	P	
Depression	Age, years	0.99	0.98	1.00	0.092	
	Gender, male	1.01	0.75	1.34	0.976	
	Secondary school or less[*]	2.07	1.49	2.88	<0.001	
	Vocational training[*]	1.88	1.31	2.70	<0.001	
	Bachelor degree[*]	1.27	0.92	1.75	0.146	
	Once or more per week[†]	0.52	0.39	0.69	<0.001	
	Less than once per week[†]	0.88	0.67	1.16	0.366	
Disability	Age, years	1.08	1.07	1.09	<0.001	
	Gender, male	1.04	0.83	1.31	0.726	
	Secondary school or less[*]	1.57	1.22	2.02	<0.001	
	Vocational training[*]	1.72	1.30	2.27	<0.001	
	Bachelor degree[*]	1.04	0.82	1.31	0.773	
	Once or more per week[†]	1.18	0.96	1.44	0.108	
	Less than once per week[†]	0.75	0.59	0.94	0.014	

Figure 7.6: Above displays the results of the effects of meditation on MS patients: ***TOP***: Meditation frequency; ***BOTTOM***: The effect of meditation on the MHC and PHC (physical health composite) (Adam B. Levin et al, 2014: *This is an open-access article distributed under the Creative Commons Attribution License, which permits unrestricted use, distribution, and reproduction in any medium, provided the original work is properly cited*).

Of 2469 eligible participants (median age was 46 years) from the USA (32.7%), Australia (25.5%), and the UK (16.9%), 52.7% of them reportedhaving meditated at least once the past 12 months. Those who meditated had a higher overall quality of life, energy, health distress, emotional well-being, health perception, and cognitive function. The mental health composite (MHC) of meditators was evidently more in participants who meditated once or more per week against those who never followed any meditation. The probability of these participants to be depressed was lower.

In a nutshell, this article described as "the first cross-sectional survey to focus solely on the association between meditation and MS health related outcomes" reported that "*participants who meditated had markedly lower depression risk. Multivariate analysis supported a significant relationship between frequency of meditation and decreased risk of depression for those who meditated once or more per week compared to those who never meditated*".

(Levin et al, 2014)

7.4 MEDITATION FOR DEPRESSION AT WORK

An Australia based randomized trial published in the *Evidence-Based Complementary and Alternative Medicine* described as "an 8-week, 3-arm, parallel randomized controlled trial which compared a "mental silence" orientated style of meditation called sahaja yoga to a "relaxation-oriented" meditative control and a waiting list (no treatment) control" was conducted at the CBD of Sydney, Australia (Trial Registration Current Controlled Trials number ISRCTN46301450). Advertisements were utilized to recruit participants to assess the effects of yoga on full-time work stress, who were employed full-time, not under any stress management approaches, non-smokers, less than 2 units of alcohol daily consumption, and no recent major life events. The intervention administered was a twice a week 1-hour evening sessions of Sahaja yoga (twice practiced at home) based on silent affirmations, breathing techniques, and attention focusing exercises for "nirvichara samadhi" or "*thoughtless awareness*". This was the mental silence meditation (MSM) group. Another relaxation control group (RC) involved participants reflecting on the day's events at the start of the meditation. The control involved subjects who were told that they were on a waiting list for subsequent admission into either meditation group. The assignment of 180 participants was random. 81% attended the maximum number of classes. Prior to treatment, mental health was assessed by GHQ28 (to assess nonspecific psychological distress and demoralization) and the Psychological Strain Questionnaire (PSQ).

The GHQ 28 mean baseline score of the sample was 7.5 suggestive of stress given that a score of 5 and more is reflective of the risk of mental health conditions. 47% of participants were at the risk of mental health morbidity. The MSM group displayed an evident improvement in the primary outcomes on the basis of intention to treat (LOCF). As against the no-treatment group, a significant improvement was displayed (P=0.546) in favor of the active intervention.

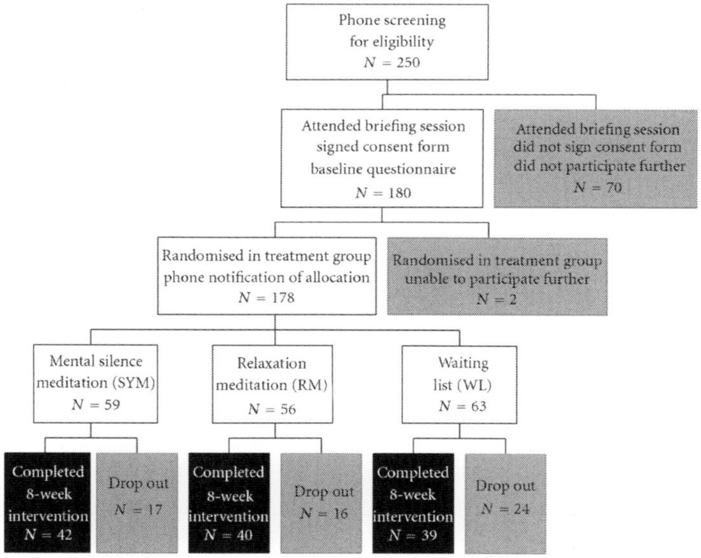

Baseline details and primary measures:

Demographic	**Mental silence**	**Relax-ation**	**No treat-ment**	P-value
Mean age (95% CI)	42.5 (39.8–45.2)	41.4 (38.9–44.0)	42.3 (39.4–45.2)	.835[a]
% White collar worker	76%	80%	64%	.123[b]
% > secondary education	46%	57%	45%	.501[b]

PSQ (95% CI)	100.5 (94.6–106.3)	100.4 (94.6–106.3)	99.9 (92.8–106.9)	.988[a]
STATE (95% CI)	41.0 (38.0–44.0)	41.3 (38.5–44.1)	40.3 (37.8–42.9)	.869[a]
DD (95% CI)	14.4 (11.2–17.6)	14.0 (12.0–17.7)	12.3 (9.8–14.8)	.384[a]

a: one-way ANOVA; b test; PSQ: Psychological Strain Inventory; STATE: State/Trait Anxiety Inventory for Adults; DD: Depression Dejection (Subscale of POMS)

Primary outcome analyses at baseline and after 8 weeks of yoga:

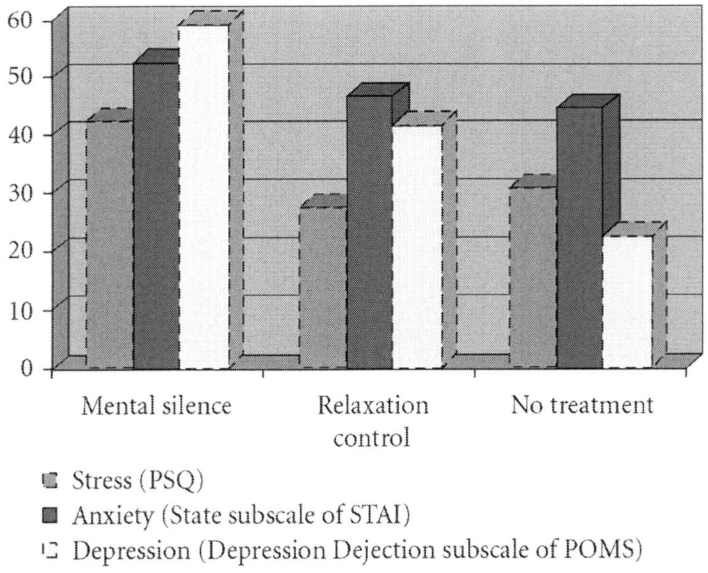

Outcome measure	Mental silence	Relaxation	No treatment	P-value
Median difference PSQ	37.0	22.3	17.5	.026[a*]
Median difference STATE	−15.0	−8.5	−9.0	.209[a]

Median difference DD	−3.0	0.0	0.0	.019[a*]

*[a]one-way ANOVA; *Significant ; PSQ: Psychological Strain Inventory; STATE: State/Trait Anxiety Inventory for Adults; DD: Depression Dejection (Subscale of POMS).*

Week 8 responses≠:

Outcome measure	Mental silence	Relxation	No treatment	*P* value
% improving 30% + in PSQ	42.4%	27.1%	30.6%	.045[a*]
% improving 30% + in STATE	52.5%	46.4%	44.4%	.651[a]
% improving 30% + in DD	59.3%	41.1%	22.2%	<.001[a**]

[a]one-way ANOVA, *Significant , **Highly significant , ≠A percentage improvement of >/=30% was classified as a positive response; PSQ: Psychological Strain Inventory; STATE: State/Trait Anxiety Inventory for Adults; DD: Depression Dejection (Subscale of POMS).

Figure 7.7: Above displays the efficacy of meditation to handle stress and depression associated with work: The scheme of the experiment, the primary outcome,and responders (R. Manocha et al, 2011: *This is an open-access article distributed under the Creative Commons Attribution License, which permits unrestricted use, distribution, and reproduction in any medium, provided the original work is properly cited*).

The team stated that "*Mental silence-orientated meditation, in this case Sahaja Yoga meditation, is a safe and effective strategy for dealing with work stress and depressive feelings. The findings suggest that "thought reduction" or "mental silence" may have specific effects relevant to work stress and hence occupational health.Given the low-cost, noncommercial nature of the intervention, and the low risk of adverse effects it would not be unreasonable to suggest that this meditation would be useful as a health enhancing strategy with potential for significant socioeconomic benefit to individuals and society".*

7.5 ADDRESSING DEPRESSION DUE TO CULTURAL ADJUSTMENTS

Research has looked at the involvement of major live events causing stress such as acculturative stress arising from adjusting to a new environment that isn't native (Berry, 2006). Given that studies have also pointed out the occurrence of this stress in Latino and Spanish-speaking populations in the US, a 2019-published randomized clinical trial looked at addressing such stress using meditation.

The study was described as "*a single-masked (rater), single-site, parallel-group, randomized controlled comparative efficacy trial of MAPs versus HE on depressive symptoms, and secondary outcomes of mindfulness and psychological stress, in moderately stressed adults who are either English- or Spanish speaking*", (ClinicalTrials.gov NCT03545074).

The participants (18 to 60 years old) were divided into those who spoke English or Spanish following advertisement, telephone screening, informed consent, and completion of questionnaires and interviews. The interventions were Mindful Awareness Practices for Daily Living Program (MAPs) and Health Education Program (HE). The former involved weekly 2-hour, 6-session group-based courses of sitting meditation, mindful eating, appreciation meditation, friendly or loving-kindness meditation, mindful walking, and mindful movement. The latter involved training on general health by a trained health educator. Assessing depression by the Beck Depression Inventory (BDI) was the primary outcome while self-reported levels of mindfulness as assessed by the Five-Facet Mindfulness Questionnaire (FFMQ) and perceived stress, as quantified by the Perceived Stress Scale were the secondary outcomes.

The assessments were completed by 71 participants: no adverse events were recorded. The severity of depressive symptoms between the start and end of the treatment was lower in those who participated in the meditation as against those who received the health education (between-treatment post-intervention mean difference: -2.2, 95% CI -4.4 –-0.07). The level of mindfulness also displayed an increase in those who meditated (between-treatment post-intervention mean difference: 10.7, 95% CI 4.5–16.9). For all participants, the improvement in depressive symptoms was 0.28 with results on similar lines for the Spanish-speaking (0.29) and English-speaking groups (0.30).

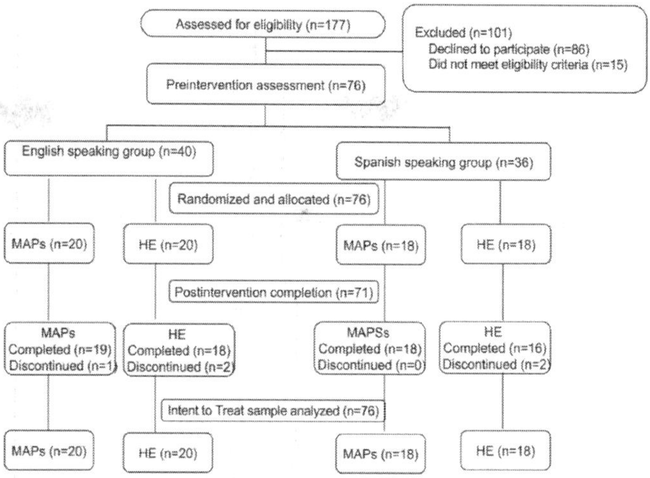

	Preintervention		
	MAPs	HE	
			Group Difference; P value
Age, mean (SD), y			
Total sample[a]	43.4 (16.4)	42.2 (14.4)	.75
English-speaking group[b]	44.2 (17.7)	43.6 (14.7)	.91
Spanish-speaking group[c]	42.5 (15.2)	40.5 (14.3)	.72
Female sex, No. (%)			
Total sample	31 (81.6)	29 (76.3)	.78
English-speaking group	15 (75.0)	16 (80.0)	1.00
Spanish-speaking group	16 (88.9)	13 (72.2)	.41
Educational level, mean (SD), y			
Total sample	16.1 (3.4)	15.9 (4.2)	.88
English-speaking group	15.7 (4.2)	16.0 (4.7)	.81
Spanish-speaking group	16.6 (2.3)	15.8 (4.0)	.52
Married, No. (%)			
Total sample	16 (43.2)	11 (29.7)	.34
English-speaking group	7 (35.0)	7 (35.0)	1.00
Spanish-speaking group	9 (52.9)	4 (23.5)	.16
Employed, No.(%)			
Total sample	21 (56.8)	27 (71.1)	.24
English-speaking group	8 (40.0)	14 (70.0)	.12
Spanish-speaking group	13 (76.5)	13 (72.2)	1.00
Beck Depression Inventory			
Total sample	17.1 (8.3)	15.9 (7.6)	.52
English-speaking group	19.6 (7.5)	17.4 (7.7)	.37
Spanish-speaking group	14.3 (8.4)	14.2 (7.3)	.99
Five Facet Mindfulness Questionnaire			
Total sample	114.6 (20.7)	119.2 (19.6)	.33
English-speaking group	108.2 (17.7)	117.0 (20.3)	.16
Spanish-speaking group	121.6 (21.9)	121.5 (19.0)	.99
Perceived Stress Scale			
Total sample	23.6 (5.5)	24.3 (5.8)	.51
English-speaking group	25.9 (5.5)	26.6 (6.8)	.71
Spanish-speaking group	20.8 (4.2)	21.8 (3.2)	.42

[a]Total sample N = 76
[b]English-speaking group N = 40
[c]Spanish speaking group N = 36

https://doi.org/10.1371/journal.pone.0219425.t001

	Preintervention (Mean SE)		Postintervention (Mean SE)		Value (95% CI)	Value (95% CI)	
	MAPs	HE	MAPs	HE	Postintervention Treatment Group Difference	Postintervention Language Group Difference[b]	
Primary Outcome							Effect Size[c]
Beck Depression Inventory							
Total sample	16.6 (0.8)	16.3 (0.8)	6.5 (0.8)*	8.7 (0.8)*	-3.2 (-4.4 to -0.07)	1.0 (-1.2 to 3.2)	.28
English-speaking group	17.4 (1.1)	16.7 (1.1)	7.0 (1.1)*	9.2 (1.1)*			.30
Spanish-speaking group	15.8 (1.1)	15.8 (1.1)	6.0 (1.1)*	8.2 (1.1)*			.29
Secondary Outcomes							
Five Facet Mindfulness Questionnaire							
Total sample	116.3 (2.2)	117.6 (2.2)	140.8 (2.2)*	130.1 (2.2)*	10.7 (4.5 to 16.9)	3.7 (-2.5 to 9.9)	.53
English-speaking group	114.2 (3.1)	116.9 (3.0)	142.0 (3.1)*	132.5 (3.0)*			.49
Spanish-speaking group	118.3 (3.2)	118.3 (3.2)	139.5 (3.2)*	127.6 (3.2)*			.59
Perceived Stress Scale							
Total sample	23.6 (0.7)	24.0 (0.7)	16.3 (0.7)*	17.6 (0.7)*	-1.3 (-3.3 to 0.7)	-2.5 (-4.6 to -0.5)	.23
English-speaking group	24.8 (1.0)	25.2 (1.0)	14.8 (1.0)*	16.6 (1.0)*			.29
Spanish-speaking group	22.4 (1.0)	22.8 (1.0)	17.8 (1.0)*	18.7 (1.0)*			.23

[a]Analyses covarying for baseline

[b]English minus Spanish score

[c]Effect size for postintervention treatment group difference

*Significant P<0.05 change from preintervention to postintervention

https://doi.org/10.1371/journal.pone.0219425.t002

Figure 7.8: Above shows the scheme of participant selection, details of the participants prior to the meditation, and the analyses of the effects of meditation on the extent of depression (Lopez-Maya et al, 2019: *This is an open-access article distributed under the terms of the Creative Commons Attribution License, which permits unrestricted use, distribution, and reproduction in any medium, provided the original author and source are credited*).

The researchers stated that "*this study is innovative by showing that benefits of mindfulness are similar to English or Spanish formats, which supports the ecological validity of mindfulness interventions in terms of language and cultural diversity. The scalability, relative low cost, and accessibility of mindfulness programs has the potential to address the need for community based interventions to be delivered to Spanish-speaking communities in the U.S. and around the world to lessen the risk of depression and possibly other adverse mental health outcomes.*" (Lopez-Maya et al, 2019).

An interesting observation is that given that such meditations can be incorporated as part of daily life with minimum hassle, apart from lowering mental depression, physiological benefits such as decreased inflammation can also be experienced. This becomes significant given that cardiovascular disorders and diabetes involve a vital involvement of depression.

7.6 TRANSCENDENTAL MEDITATION®

Transcendental Meditation® (TM)introduced in the West by Maharishi Mahesh Yogiis described as an approach for meditation and relaxation. Early research in 1997 assessed the benefits of this approach on depressive symptoms. 44 federal government agency employees were randomly assigned to either 12 weeks of Corporate Stress Management or Transcendental Meditation®. Tests after 3 months of the meditation revealed a lowered profile of depression and anxiety. The meditators after 3 years also demonstrated lowered depression.

(Sheppard et al, 1997; Elder et al, 2014)

A study published in 2008 in *Clinical practice and epidemiology in mental health* by researchers Yunesian and team assessed the effects of TM on the mental health of adults in Tehran (Iraq: according to the authors, the extent of mental distress was 21%, as assessed by the Persian version of the GHQ-28 questionnaire).

85 participants were recruited from certified yoga clubs of Tehran and subjected to a 12-week course from November 2005 until January 2006. Following initial intensive training, participants received a personal mantra on the basis of their traits on which they meditated for 15 – 20 minutes. Assessment by the 28-item General Health Questionnaire (GHQ-28) revealed that at the time of enrolment, mental distress was seen in 38.8% of the participants that dropped to 26.3% following the 12-week study. Analyses were suggestive of no role of age, sex, and marital status on this improvement.

Table 7.1: below shows how the mantra-based meditation lowered the mental distress in the clinical study(Yunesian et al, 2008; *This is an Open Access article distributed under the terms of the Creative Commons Attribution License (http://creativecommons.org/licenses/by/2.0), which permits unrestricted use, distribution, and reproduction in any medium, provided the original work is properly cited*)

* *All p values are for Wilcoxon rank score test*

	Subgroup n (%)	**GHQ ≥ 22 n (%)**	**Mean GHQ subclass scores (before – after)**				**GHQ-28 score**
			Somatisation	***Anxiety***	***Social Dysfunction***	***Depression***	
Total	80	31 (39%)	(6.2 – 4.8) p* < 0.001	(6.6 – 4.7) p < 0.001	(9.2 – 9.4) p = 0.979	(3.0 – 2.0) p = 0.152	(25.1 – 21.3) p < 0.001
Age Group							
20 – 29	35 (44%)	13 (37%)	(6.1 – 4.9) p < 0.001	(6.8 – 5.5) p < 0.001	(8.9 – 8.9) p = 0.979	(3.9 – 3.0) p = 0.152	(25.7 – 22.3) p < 0.001
30 – 39	25 (31%)	10 (40%)	(7.5 – 5.5) p = 0.019	(7.7 – 4.7) p = 0.006	(8.5 – 10.1) p = 0.352	(3.5 – 1.9) p = 0.106	(27.2 – 22.4) p = 0.028
≥ 40	20 (25%)	8 (40%)	(6.0 – 4.6) p = 0.023	(6.6 – 3.8) p = 0.036	(10.4 – 9.8) p = 0.292	(1.8 – 1.5) p = 0.733	(24.9 – 19.7) p = 0.022
Gender							

Male	24 (30%)	7 (29%)	(5.6 – 3.8) p = 0.005	(5.3 – 3.7) p = 0.029	(8.5 – 9.2) p = 0.667	(2.4 – 1.5) p = 0.094	(21.9 – 18.3) p = 0.01
Female	56 (70%)	24 (43%)	(6.5 – 5.3) p = 0.018	(7.2 – 5.1) p < 0.001	(9.5 – 9.6) p = 0.839	(3.3 – 2.7) p = 0.517	(26.4 – 22.7) p = 0.003
Marital Status							
Married	44 (55%)	17 (39%)	(5.8 – 4.7) p = 0.025	(6.1 – 4.1) p = 0.01	(9.3 – 9.8) p = 0.793	(2.7 – 2.0) p = 0.253	(23.8 – 20.6) p = 0.024
Unmarried	36 (45%)	14 (39%)	(6.8 – 5.1) p = 0.006	(7.3 – 5.5) p = 0.003	(9.1 – 9.0) p = 0.742	(3.5 – 2.7) p = 0.350	(26.6 – 22.3) p < 0.001

It was concluded by the authors that "*Meditation improves mental health of volunteer yoga club members and the effects are mostly on relieving somatisation and anxiety. The favourable outcomes of meditation appear to be independent of age, sex and marital status.*"
(Yunesian et al, 2008)

7.7 MANTRAS SHOW THE WAY

Mantra is a mystic syllable with esoteric meaning. The vibrations attributed to mantras have been implicated to cause changes that help alleviate stress manifestations inclusive of depression (Sharma, 2018).

Researchers Lynch and team (*European Journal of Integrative Medicine*, 2018) state "*Mantra meditation typically involves continually repeating a chosen word, phrase or set of syllables (silently or aloud) while passively disregarding any internal or external distractions. The sound or mantra in meditation is proposed to act as an effective vehicle for overriding mental speech (which is the predominant form of conscious thinking for most people), thus continually redirecting negative or intrusive automatic thoughts which perpetuate psychological distress*".

"One who is engaged in the practice of concentration, uttering the monosyllable *Om* (the *Brahman* or consciousness) who remembers it always, he attains the supreme goal"-*Bhagavad Gita*

The combination of A, U, and Myield "OM"-considered the syllable of the past, the present, and the future, according to the *Mandukya Upanishad*. The classical yoga text *Patanjali's Yoga Sutras* describe OM as *Tasyavachakahpranavah* (Ch: I; V: 27) or that Om is the Divine or what is called as God.

(Kumar et al, 2010)

Early work in 1995 showed that the heart rate was evidently lowered in seven experienced *Om* meditators during the OM meditation against a control period of no thinking (Telles et al, 1995).

Analyses of the middle latency auditory evoked potentials in experienced OM meditators revealed that there were neural changes with a significant increase in the peak amplitude of the Na wave when OM was meditated upon while a dip was seen when the neutral word "One" was meditated upon. This becomes significant as researchers Kumar and team (2010) state "Scientific studies on *Om* suggest that the mental repetition of *Om* results in a physiological state at one time characterized by reduced physiological alertness, increased sensitivity as well as synchronicity, as well as changes at specific levels along the auditory pathway suggestive of increased sensitivity to sensory transmission."

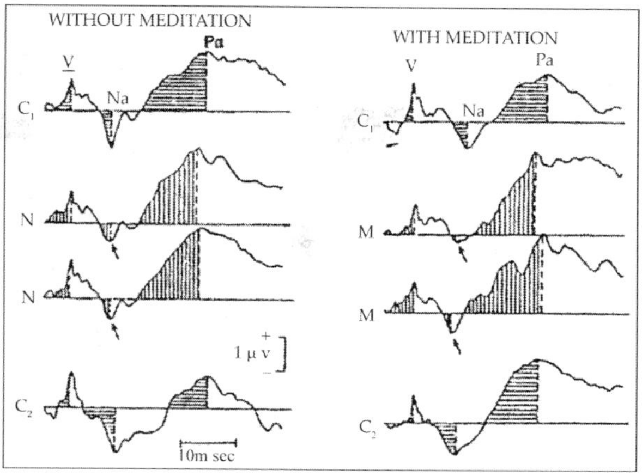

Figure 7.9: Above shows the middle latency auditory-evoked potentials (AEP-MLRs)on the left (a meditator with 12 years of practice) and right (meditator since 10 days) showing the neuronal changes due to OM chanting or meditation (Kumar et al, 2010; *From Telles et al. 1994, Internal J. Neurosci: This is an open-access article distributed under the terms of the Creative Commons Attribution License, which permits unrestricted use, distribution, and reproduction in any medium, provided the original work is properly cited*).

Researchers Kalyani and team published in the *International journal of* yoga that the effect of OM chanting was on the same lines of vagus nerve stimulation that finds use in addressing depression. 12 healthy literate volunteers, 22-39 years old (9 male) were assessed to be devoid of substance abuse or no major psychiatric disorder received training to chant 'OM' by an experienced yoga teacher. The chanting was loud: the vowel (O) part of the 'OM' for 5 sec continuing into the consonant (M) part of the 'OM' for the next 10 sec. The control was 15 seconds of chanting'sssss….' to account for the expiration of OM.

MRI imaging and analyses revealed that while there was no part of the brain activated by the OM chanting, the amygdala, anterior cingulate gyrus, hippocampus, insula, orbitofrontal cortex, parahippocampal gyrus, and thalamus were conspicuously deactivated. No such observation was seen in the control.

15 Seconds	'OM'	15 Seconds	SSSS
REST	15 Seconds	REST	15 Seconds

Brain region**	X	Y	Z	T	FWE-p*
Right amygdala	24	-10	-08	5.2	<0.001
Left anterior cingulate gyrus	-02	45	-02	10.2	<0.001
Right anterior cingulate gyrus	12	49	-01	9.8	<0.001
Left hippocampus	-32	-18	-11	6.5	<0.001
Right hippocampus	30	-31	-05	4.6	<0.001
Left insula	-28	19	-06	6.5	<0.001
Right insula	38	15	-06	4.9	<0.001
Left orbitofrontal cortex	-28	29	-08	6.6	<0.001
Right orbitofrontal cortex	30	29	-08	7.3	<0.001
Left parahippocampal gyrus	-30	-20	-21	5.1	<0.001
Right parahippocampal gyrus	32	-28	-22	5.0	<0.001
Left thalamus	-14	-05	13	6.6	<0.001
Right thalamus	16	-07	11	6.2	<0.001

***X, Y, Z Talairach coordinates of peak activation*

**Family-wise error corrected 'P'* value

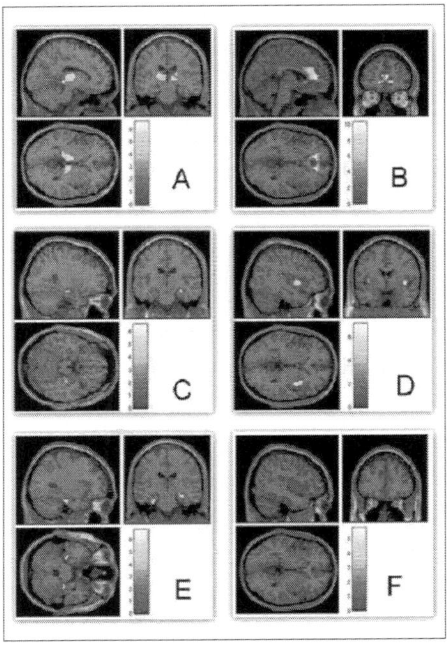

Figure 7.10: Above shows the scheme of OM chanting followed by the tabular representation of which brain portions were deactivated by OM chanting and the scanshowing the deactivation of thalami (A) and limbic structures - anterior cingulum (B), hippocampi (C), insula (D) and parahippocampi (E);T scores

given in the table are shown in the color bar (Kalyani et al, 2011: *This is an open-access article distributed under the terms of the Creative Commons Attribution License, which permits unrestricted use, distribution, and reproduction in any medium, provided the original work is properly cited*).

The deactivation of the amygdala, parahippocampal, hippocampal regions observed in this work was on the same lines of vagus nerve stimulation (VNS). Lowered blood flow in the hippocampus, amygdala, and cingulate gyri was detected by positron emission tomography in VNS in epilepsy patients while VNS also lowered blood flow in the amygdala, left hippocampus, left subgenual cingulate cortex, left and right ventral anterior cingulum, right thalamus and brainstem-this becomes significant given that depression shows hyperactivation of these regions.

Researchers Das and Anand (*Psychological Thought*: 2012) assessed the effect of OM chanting on the galvanic skin response (GSR)of20 students of the Psychology Department of Dayalbagh Educational Institute, Agra. The GSR refers to atransient change in certain electrical properties of the skin sweat glands due to any stimulus. As an indicator of the autonomic nerve response, the GSR can indicate stress.

18 to 22-year-old female participants participated in a 3-day 30-minute prayer and meditation session (15 minutes each).

Prayer: *"Fold your hands, close your eyes and think about your God. Now concentrate on background musical prayer.*"

Meditation: *"First, enter your meditative state by securing a quiet, dimly lit comfortable sitting position. Background meditative music will help facilitate migration into deeper brain centers. Eyes can be open or shut. Breathe comfortably and slowly through your nose, into your belly and then into your chest. Feel that positive energy is running into all over body and each part of your body is pure and holy. Slowly let the air exhale through your mouth. Now chant the mantra, "OM" Visualize a situation when you were very happy and feel that happiness. Feel the positive energy and enjoy in this positive environment"*.

Given that GSR decreases with stress, the study recorded an evident increase of the GSR in the participantsfrom (Mean- 388.82, SD- 353.64) to (Mean-817.53, SD-449.83) following the chanting and meditation. The researchers concluded that chanting mantras and meditation leads to psychophysiological relaxation, vital to handle depressive stress.

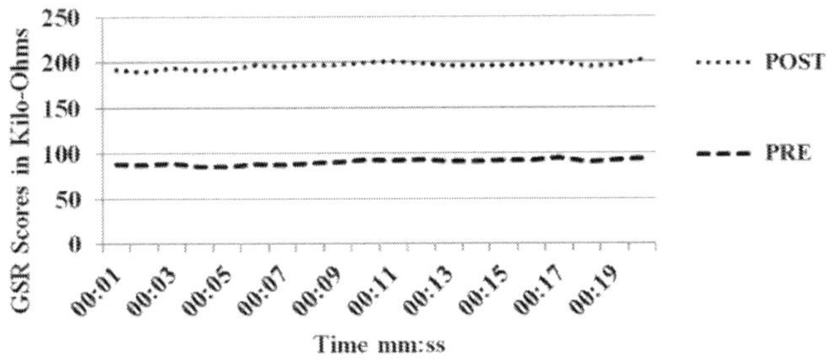

Figure 1. Graph showing data measured on a single subject

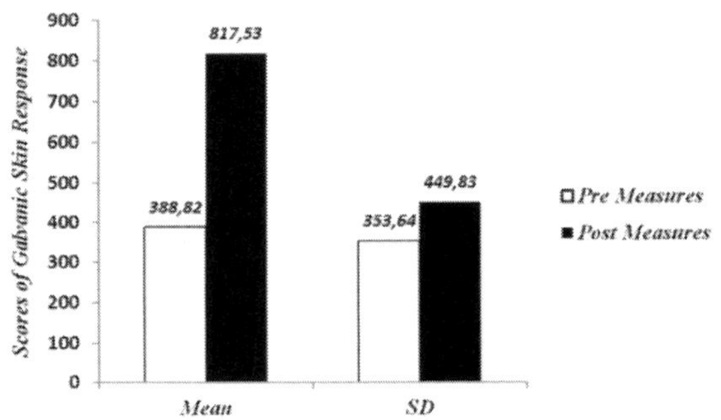

Figure 3. Comparison between pre-intervention GSR scores and post-intervention GSR scores

Figure 7.11: Above displays the changes in the GSR: ***TOP***: single subject measurement; ***BOTTOM***: pre-intervention and post-intervention GSR scores (Das and Anand, 2012: *This is an open-access article distributed under the terms of the Creative Commons Attribution License (http://creativecommons.org/licenses/by/3.0), which permits unrestricted use, distribution, and reproduction in any medium, provided the original work is properly cited*).

Table 7.2: Other studies exploring the effects of OM (Harne et al, 2019; *This is an open-access article distributed under the terms of the Creative Commons Attribution License (https://creativecommons.org/licenses/by/4.0), which permits unrestricted use, distribution, and reproduction in any medium, provided the original work is properly cited*)

Om chanting type	Subjects	Type of meditators	Method / Signal Processing	Experimental Design	Findings
Loud Om chanting	20	naïve	Functional near-infrared spectroscopy (fNIRS)	Subject exposed to mobile phones (30 min) → Om chanting (5 min) Study suggests that Om chanting may have a stimulating effect on branch of vagus nerve in the ear canal.	Deactivation of pre-frontal cortices due to the vibrations produced by the sound 'Om';
Listening Om mantra	21 (All male)	-	Functional magnetic resonance imaging (fMRI)	Listening OM (12 ms) → No sound (12 ms) → Listening AAM (12 ms)→ No sound (12 ms) → Listening TOM (12 ms) → No sound (12 ms) The common activated region DMFC supports the emotional empathy of "Om" sound, while SMG implicates phonological processing of "Om" syllable.	Neural regions activated during listening to "Om" sound in contrast to non-meaningful word (TOM);

Mental Om mediation	22 (All male)		Experienced	fNIRS	Random thinking → mental chanting and effortless defocusing on syllable "Om"	Oxygenation levels are increased in the PFC during meditation
Loud Om mantra chanting	10	Naïve	Time domain Analysis (Higuchi Fractal Dimension (HFD))		Rest → Om Chanting → Rest	Om chanting reduces the complexity of EEG signal.

According to Sharma (2018), Gayatri mantra chanting augments the quality known as "sattva" or light and mental clarity thereby diminishing the effects of stress such as depression.

7.8 SUDARSHAN KRIYA

Sudarshan Kriya yoga is a breathing-based meditation approach that is receiving attention in research studies. For instance, a controlled pilot trial was published in *BMC complementary and alternative medicine* by researchers Kjellgren and team assessing whether Sudarshan Kriya and related practices (SK&P) could augment "feeling good" in volunteers(Australian Clinical Trial Registry ACTRN012607000175471). Of 103 participants, 48 participants (13 men, 35 females: mean age of 31.95 years, SD = 11.85) participated in SK&P at the Art of Living Yoga Center in Karlstad and Eskilstuna, Sweden while 55 control participants (12 men, 43 females; mean age of 31.90 years, SD = 10.84) were yoga-interested students at Karlstad University.

Sudarshan Kriya (SK) and related practices (SK&P) involved in the following order done with closed eyes (a) Three-Stage Pranayama with Ujjayi or "Victory Breath", (b) three sets of Bhastrika or "Bellow's Breath", and (c) SK or the "Healing Breath Technique". The first step involves placing the arms at particular locations and breathing in Ujjayi breath (2 to 4 breaths per minute vs. the regular 14 to 16 breaths per minute). In Bhastrika, fast and vigorous breath patterns are followed with arm movements. SK involves breathing in a cyclic format: 20 slow respiratory cycles per minute followed by 40–50 medium respiratory cycles per minute, and fast 60–80 cycles per minute. The controls were asked to focus on their breath for 15 minutes with their eyes closed. The outcomes assessed were depression, anxiety, mood, optimism, energy levels, and experience of altered states of consciousness as assessed by the HAD – Hospital Anxiety Depression Scale, LOT – Life Orientation Test, SE – Stress and Energy, PANAS – Positive affect and Negative Affect Scales, PAI – Pain Area Inventory, and EDN – Experienced deviation from the normal state.

Analyses revealed that the practice of SK resulted in the depression scores to drop from 3.67 (*SD* = 2.56) to 2.97 (*SD* = 2.02) while the level of optimism rose from 21,29 (*SD* = 5.27) at pretest to 22.60 (*SD* = 5.16) at posttest. The HAD scores showed that depression was lowered by 33% in those who practiced the breathing-based meditation. In the words of the researchers, "*These easily learned and applied cost-effective yogic practices can thus be offered as an intervention to the adult population at large to relieve psychosocial stress and its associated disorders; this in turn may result in the prevention of a score of mental and physical disorders. Full-scale trials are required to conclusively assess the validity of this hypothesis*" (Kjellgren et al, 2007).

Table 7.3: Below displays the baseline features of the participants (Kjellgren et al, 2007; *licensee BioMed Central Ltd; This is an Open Access article distributed under the terms of the Creative Commons Attribution License (http://creativecommons.org/licenses/by/2.0), which permits unrestricted use, distribution, and reproduction in any medium, provided the original work is properly cited*): 'Sleep Hours' refers to the number of daily sleep, 'Min Awake' the number of minutes before falling asleep, 'Sleep Quality' refers to perceived quality of sleep, 'cig' are the number of cigarettes per month, 'Alc' is alcohol. The numbers in parentheses are standard deviation

	SK&P	**Controls**
Age	31.95 (11.85)	31.90 (10.84)
Depression	4.07 (3.078)	3.25 (2.026)
Stress	2.68 (0.93)	2.61 (0.92)
Energy	2.99 (0.91)	2.97 (0.77)
Sleep Hours	7.46 (0.84)	7.68 (0.50)
Min Awake	20.12 (18.79)	17.71 (13.27)
Sleep Quality	64.82 (21.37)	70.28 (17.94)
cig/month	12.47 (41.61)	15.00 (55.46)
snuff/month	1.40 (7,48)	0.13 (0.53)
Alc ml/month	184.915 (173.83)	156.73 (143.47)
Optimism	21.84 (5.72)	24.88 (3.86)
Anxiety	7.58 (3.58)	4.63 (2.59)

In the ongoing quest to address the morbid suicides associated with posttraumatic stress disorder (PTSD) in veterans from the Operation Enduring Freedom (Afghanistan war) or Operation Iraqi Freedom (Iraq War),

Seppälä and team (2014) assessed the utility of breathing-based meditation in a randomized controlled longitudinal study. 21 veterans who fulfilled the criteria were randomly assigned into Sudarshan Kriya yoga active (n = 11) or waitlist control (n = 10): ages were for the active: M = 28.09 years, SD = 2.91; control: M = 29.20, SD = 6.66. The Combat Exposure Scale revealed that the nature and severity of exposure to traumatic combat experiences were similar for both groups. The intervention was for 7-days: net 21-hour intervention. Assessments were conducted using the PTSD Checklist-Military version and the Mood and Anxiety Symptoms Questionnaire (MASQ) along with startle response measurements.

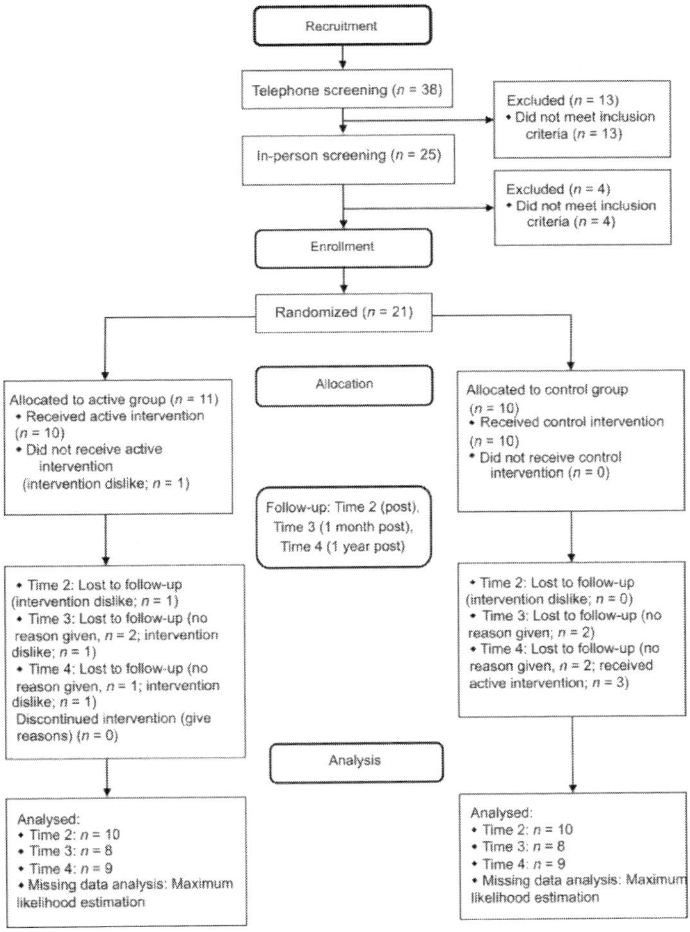

Figure 7.12: Above shows the scheme of recruiting war veterans (Seppälä et al, 2014; *The Authors. Journal of Traumatic Stress published by Wiley Subscription Services, Inc. on behalf of International Society for Traumatic Stress Stud-*

ies. This is an open-access article under the terms of the Creative Commons Attribution License, which permits use, distribution and reproduction in any medium, provided the original work is properly cited).

The time-frames were Time-1: start; Time-2: 21 hours; Time 3: 1-month post-intervention, and Time 4: 1-year post-intervention, with no adverse effects reported. Lesser PSTD symptoms were reported at Time 2, Time-3 and Time-4 by the meditators. Re-experiencing was lower at Time 2, $t(45.28) = 2.11$, $p = .040$, and Time 4, $t(45.36) = 3.13$, $p = .003$, while the values for Hyperarousal were lesser at Time 2, $t(44.88) = 5.45$, $p < .001$, Time 3, $t(45) = 4.53$, $p < .001$, and Time 4, $t(44.93) = 5.06$, $p < .001$, against Time 1.

Table 7.4: Below displays the results of the analyses at the 4 time-points in the effects of breathing-based meditation on war veterans (Seppälä et al, 2014; *The Authors. Journal of Traumatic Stress published by Wiley Subscription Services, Inc. on behalf of International Society for Traumatic Stress Studies. This is an open-access article under the terms of the Creative Commons Attribution License, which permits use, distribution and reproduction in any medium, provided the original work is properly cited*)

Key: PCL-M = Posttraumatic Stress Disorder Checklist-Military Version; MASQ = Mood and Anxiety Symptoms Questionnaire; GDA = General Distress-Anxiety; AA = Anxious Arousal; GDD = General Distress-Depressive; AD = Anhedonic Depression.

[a]Time 1 has n = 10 participants due to technical difficulties.

	Time Control							
	1 (n = 10)		**2 (n = 10)**		**3 (n = 8)**		**4 (n = 5)**	
Measure	***M***	***SD***	***M***	***SD***	***M***	***SD***	***M***	***SD***
PCL-M								
Reexperiencing	8.00	4.16	8.60	5.21	7.63	5.07	10.00	5.34
Avoidance	13.40	5.97	12.80	4.73	11.50	4.17	15.00	4.53
Hyperarousal	11.00	4.69	11.10	5.80	10.63	6.44	13.80	6.61
Total	32.40	13.34	32.50	15.01	29.75	15.24	38.80	15.12
MASQ								

GDA	15.90	3.98	17.30	4.11	19.25	6.14	22.40	4.16
AA	21.30	3.65	20.60	2.59	22.25	6.25	25.80	7.05
GDD	21.60	7.34	19.40	4.79	20.13	6.81	24.20	6.53
AD	66.10	12.68	64.10	13.07	62.63	12.37	68.00	12.39
Total	124.90	17.51	121.40	19.40	124.25	22.41	140.40	18.37
Startle								
Light	32.26	36.80	29.24	27.16				
Dark	35.08	32.02	31.46	27.46				
Respiration	10.03	2.05	11.16	2.23				

	Time Active							
	1 (*n* = 11)		**2 (*n* = 10)**		**3 (*n* = 8)**		**4 (*n* = 9)**	
Measure	***M***	***SD***	***M***	***SD***	***M***	***SD***	***M***	***SD***
PCL-M								
Reexperiencing	9.27	3.23	7.70	2.26	8.88	3.72	7.11	1.69
Avoidance	14.45	6.52	10.10	3.14	10.50	4.41	10.22	3.83
Hyperarousal	12.82	4.71	8.20	2.53	9.25	3.73	8.22	2.59
Total	36.55	11.44	26.00	6.91	28.63	10.68	25.56	6.58
MASQ								
GDA	19.64	5.50	16.00	4.27	16.00	4.41	18.44	6.41
AA	27.00	7.84	22.70	4.74	20.75	2.49	21.67	6.40
GDD	23.00	8.04	18.10	5.17	17.38	5.37	18.78	7.82
AD	59.64	17.23	44.60	15.36	49.00	12.62	47.11	17.60
Total	129.27	32.47	101.40	26.60	103.13	22.65	106.00	35.96
Startle								
Light	82.04	104.45	61.37	68.95				
Dark	94.40	103.08	66.11	65.12				
Respiration[a]	10.20	2.12	9.07	2.81				

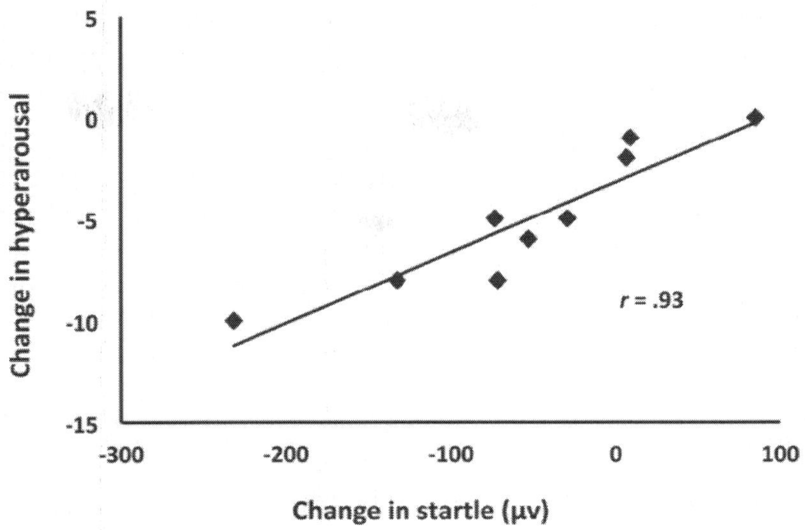

Figure 7.13: Above shows the diminishing of hyperarousal symptoms of PSTD following Sudarshan Kriya in war veterans i.e. lowered startle response (Seppälä et al, 2014; *The Authors. Journal of Traumatic Stress published by Wiley Subscription Services, Inc. on behalf of International Society for Traumatic Stress Studies. This is an open-access article under the terms of the Creative Commons Attribution License, which permits use, distribution and reproduction in any medium, provided the original work is properly cited*).

Given that there are challenges remaining to be addressed with conventional medication, this report brings hope for the survivors of trauma. According to the authors "*Sudarshan Kriya yoga, a week-long intervention with longitudinal benefits, shows promise as a viable alternative or adjunct intervention for addressing PTSD and suicide in returning veterans*".

According to Zope and Zope (2013), depression and anxiety associated with other disasters such as the New York World Trade Center 9/11, the 2004 Southeast Asia tsunami, the 2000 Gujrat earthquake, and Hurricane Katrina were relieved by SKY (Sudarshan Kriya Yoga)-breathing meditation. The P300 event-related potential (ERP) amplitude of an electroencephalogram (EEG) that measures an abnormality in the brainwave in depressed people reached the normal level of health controls by day 90 of SKY.

"*SKY practices are cost-effective, well-tolerated tools that can be easily integrated into diverse community care models. SKY relieves stress and develops an individual's mind–body–spirit so that they can be happier, healthier, and possibly even longer lived. In the competitive modern world, in which stress and anxiety are part of everyday life, adding a time-honored, evidence-based breathing program like SKY may facilitate a healthy life*"- Zope and Zope (2013).

7.9 FINAL WORDS

According to Deshmukh (2006), "*Dhyana-Yoga* is a Sanskrit word for the ancient discipline of meditation, as a means to *Samadhi* or enlightenment. Samadhi is a self-absorptive, adaptive state with realization of one's being in harmony with reality. It is unitive, undifferentiated, reality-consciousness, an essential being, which can only be experienced by spontaneous intuition and self-understanding. Meditation is an art of being serene and alert in the present moment, instead of constantly struggling to change or to become. It is an art of efficient management of attentional energy with total engagement (*poornata*, presence, mindfulness) or disengagement (*shunyata*, silence, emptiness). In both states, there is an experience of spontaneous unity with no sense of situational interactive self or personal time. It is a simultaneous, participatory consciousness rather than a dualistic, sequential attentiveness. There is a natural sense of well being with self-understanding, spontaneous joy, serenity, freedom, and self-fulfillment. It is where the ultimate pursuit of happiness and the search for meaning of life resolve. One realizes the truth of one's harmonious being in nature and nature in oneself. It is being alive at its fullest, when each conscious moment becomes a dynamic process of discovery and continuous learning of the ever-new unfolding reality".

Apart from what people feel, research too is highlighting the health benefits of meditating. Apart from augmenting the well-being, mindfulness-based interventions (MBIs) are emerging to be promising to address mental health issues including depression.

A research article published by Tran et al (2014) looked at the effects of meditation in 2 populations: German and Spanish on stress, depression, and the factors by which meditation addressed mindfulness. Participants of the Spanish group practiced more Zen and Vipassana while the German group reported Tibetan meditation, yoga, Transcendental Meditation, Naikan, Qi Gong, Tai Chi, and autogenic training. Mindfulness was gauged by the Five Facets Mindfulness Questionnaire (FFMQ): the 5 aspects of the FFMQ are Observe, Describe, Act with Awareness, Nonjudging of Inner Experience, and Nonreactivity to Inner Experience. Depression was assessed by the Brief Symptom Inventory (BSI) in the German sample and the Depression, Anxiety and Stress scale (DAS-21) in the Spanish sample.

Decentering and Nonattachment were the prominent mechanisms by which mindfulness exerted its benefits. The least factor associated with mindfulness was Emotional Awareness and Activation Control.

	German Sample	Spanish Sample	Statistical test
n	891	393	
Women, *n* (%)	661 (74.2%)	220 (56.0%)	$\chi^2(1) = 41.98^{***}$
Age, *M* (*SD*)	48.65 (10.52)	43.97 (10.60)	$t(1280) = 7.30^{***}$, $d = 0.44$
Min–Max	20–82	18–75	
Education			
Primary or lower secondary	10 (1.1%)	7 (1.8%)	$\chi^2(2) = 54.26^{***}$
Upper secondary	363 (40.7%)	77 (19.6%)	
University/diploma	518 (58.1%)	309 (78.6%)	
Frequency meditating			
> Once a week	675 (75.8%)	256 (66.1%)[a]	$\chi^2(2) = 13.46^{***}$
Once a week	66 (7.4%)	46 (11.9%)[a]	
< Once a week	150 (16.8%)	85 (22.0%)[a]	
Meditation experience in years (*Md*)[b]	8.00	4.00	*U* test, $z = 7.90^{***}$
Min–Max, interquartile range	0–45, 4–15	0–43, 2–10	
Meditation type[c]			
Zen	102 (11.5%)	110 (39.0%)	$\chi^2(4) = 369.32^{***}$
Vipassana	107 (12.0%)	136 (48.2%)	
Tibetan	54 (6.1%)	5 (1.8%)	
Yoga	348 (39.1%)	24 (8.5%)	
Other	279 (31.3%)	7 (2.5%)	

Note. [a] Based on 387 participants due to incomplete data. Based on [b] 738 and 348/[c] 890 and 282 participants, respectively, for which data were available. *** $p < .001$.
doi:10.1371/journal.pone.0110192.t001

Figure 7.14: Aove shows the sample features followed by correlating the meditation experiences and depression and finally the results in table and figure formats showing the pathways associated with mindfulness (Tran et al, 2014:*This is an open-access article distributed under the terms of the Creative Commons Attribution License, which permits unrestricted use, distribution, and reproduction in any medium, provided the original author and source are credited*).

Correlations of FFMQ Factor Scores (Short Form) with Meditation Experience, Depression, and Anxiety in Both Samples, and Means and Standard Deviations of Factor Scores and of Depression and Anxiety, Differentiated By Sample

	Correlation with			***M* (*SD*)**	
	Med. exp.[a]	**Depression[b]**	**Anxiety[b]**	**German**	**Spanish**
Observe	.20	-.30	-.30	0.17 (0.93)	-0.01 (0.87)
Describe	.18	-.31	-.26	0.02 (0.87)	-0.02 (0.84)
Actaware	.28	-.40	-.41	0.42 (0.89)	0.00 (0.85)
Non-judge	.24	-.42	-.41	0.18 (0.90)	-0.04 (0.86)
Non-react	.29	-.45	-.44	-0.08 (0.98)	0.00 (0.86)
SRA	.29	-.46	-.45	-0.07 (1.10)	0.00 (0.97)
OTE	.33	-.45	-.46	0.55 (0.95)	0.00 (0.90)
Depression[b]	-.11	–	.58	0.46 (0.54)	6.14 (7.70)
Anxiety[b]	-.16	.58	–	0.49 (0.51)	5.14 (6.27)

Note: Med. exp. = mediation experience; SRA = Self-regulated Attention; OTE = Orientation to Experience. [a] Log-months were used for correlational analyses (n = 1086 due to incomplete data). [b] Measured with the BSI in the German sample and the DAS-21 in the Spanish sample; standardized scores were used for correlational analyses, means and standard deviations refer to raw scores (n = 367 in the Spanish sample due to incomplete data). For correlation coefficients, all ps< .001.

	Standardized path coefficients			Total effect of meditation
	SRA	OTE	R^2	experience
Decentering	.45	.40	61%	.25
Nonattachment	.39	.36	47%	.22
Acceptance of Emotions[a]	.21	.46	38%	.19
Control & Regulation[a]	.25	.41	37%	.19
Emotional Clarity[a]	.23	.40	35%	.18
Attentional Control	–	.59	35%	.15
Body Awareness	.52	-	27%	.17
Inhibitory Control	.35	.21	26%	.17
Goals[a]	–	.51	26%	.14
Body Association[b]	–	.33	11%	.09
Emotional Awareness[a]	.33	-	10%	.10
Activation Control	–	.32	10%	.08

Note. Results of a path model on the effects of meditation experience on mindfulness and mechanisms of mindfulness in the Spanish data. SRA = Self-regulated Attention; OTE = Orientation to Experience. [a] DERS scales were reverse scored so that higher scores reflected greater ability, not more problems with emotion regulation. [b] The Body Dissociation scale was reverse scored so that higher scores reflected bodily association. All $ps<.001$, except the path of SRA on Control & Regulation, $p=.002$.

doi:10.1371/journal.pone.0110192.t007

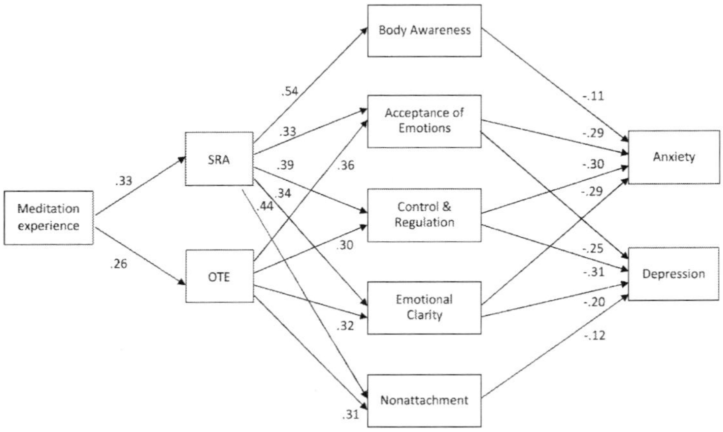

Figure 7.15: Important mechanisms associated with handling depression emerged as Acceptance of Emotions and Nonattachment. Interestingly, in the case of anxiety, the vital factors were Emotional Clarity and Body Awareness. This has been represented in the scheme (Tran et al, 2014: *This is an open-access article distributed under the terms of the Creative Commons Attribution*

License, which permits unrestricted use, distribution, and reproduction in any medium, provided the original author and source are credited): *SRA = Self-regulated Attention; OTE = Orientation to Experience.*

Analyzing the psychological factors such as Acceptance of Emotions and Nonattachment in alleviating depression opens up the avenue of comprehending the details of therapeutic approaches based on meditation and mindfulness.

Given the burgeoning number of studies and attention meditation is receiving, a cross-sectional online survey discussing new insights into the aspects of meditation was reported by a task force of meditation researchers and teachers:Vieten and team (2018). Analyses of 1,130 participants from the United States (57%), Canada (8%), United Kingdom (8%), Australia (4%), India (2%), Portugal (2%), Germany (2%), and New Zealand, Norway, and Mexico (1%), with the rest from other nations, revealed that 19% of them suffered from depression. In terms of the number of years of meditation practices, regular meditation (at least once a week) accounted for 14.7 ± 13.5 (range 0–75) years! The most practiced form was breath-focused meditation followed by mindfulness, Transcendental Meditation, body scan, contemplative prayer, mantra chanting, and visualization. More than 40% of the participants reported mystical experiences, as assessed by an adapted version of the Revised Mystical Experience Questionnaire (MEQ30).

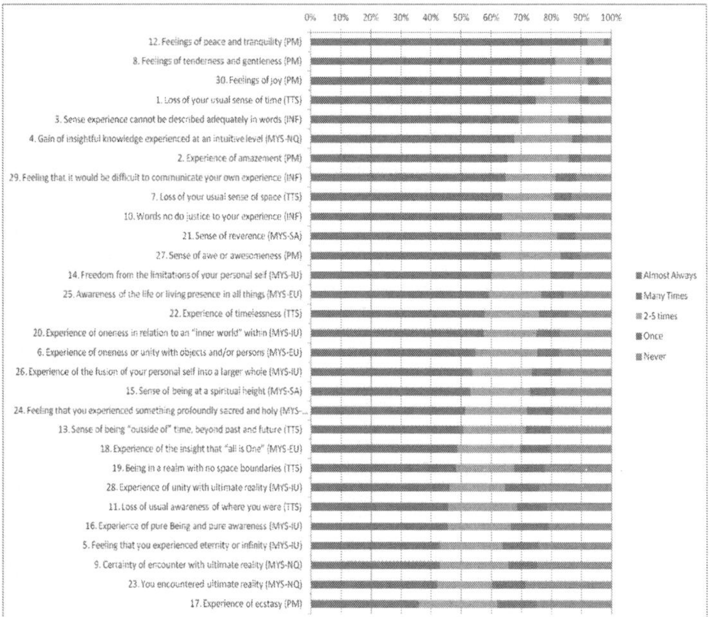

	No	Yes	Frequency: One Time	2–5 Times	Many Times	Almost Always
Physical						
Altered breathing	12	88	2	12	42	33
Sensations in body	15	85	4	18	46	17
Altered hearing	20	80	7	23	39	12
Altered vision	28	72	6	22	33	10
Altered body sensations	39	61	5	23	26	7
Altered smell/taste	65	35	5	18	11	1
Anomalous Spatial-Temporal						
Altered time	14	86	4	19	44	18
Altered space	40	60	6	21	27	7
Increased synchronicities	18	82	2	15	48	17
Nonphysical entities	40	60	8	20	26	6
Cognitive/Psychological						
Altered awareness	9	91	6	18	49	18
Aha! Moments	11	89	4	24	53	9
Altered identity	31	69	8	20	33	9
Disturbing emotions	68	32	9	16	7	0
Relational						
Collective energy	24	76	6	23	39	8
Connection with teacher	47	53	8	17	23	5
Extended Phenomena						
Clairvoyance/ Telepathy	44	56	6	20	24	6
External Physical Phenomena	69	31	6	16	9	0

Percentage endorsed of each item ordered by most to least common within domain. No = this has never happened. Yes = sum of remaining responses (This has happened once, This has happened 2–5 times, This has happened many times, This almost always happens). Frequency details the percentage of participants who reported each frequency. Full wording for each item is available in online supplemental materials. N = 1,120

https://doi.org/10.1371/journal.pone.0205740.t003

Figure 7.16: Above displays the ***TOP***: beneficial; ***BOTTOM***: mystical/ extraordinary events associated with meditation as surveyed by meditation teachers and researchers (Vieten et al, 2018: *This is an open-access article distributed under the terms of the Creative Commons Attribution License, which permits unrestricted use, distribution, and reproduction in any medium, provided the original author and source are credited*).

These aspects open up new avenues of research in meditation and mindfulness. To quote the researchers, "*The intention of this paper is to invite all of us to step into a new paradigm from which to explore one of the greatest of human quests—the understanding, healing, and enhancement of the human mind*" (Vieten et al, 2018).

"*Meditation will not carry you to another world, but it will reveal the most profound and awesome dimensions of the world in which you already live. Calmly contemplating these dimensions and bringing them into the service of compassion and kindness is the right way to make rapid gains in meditation as well as in life.*"

– Zen Master Hsing Yun

7.10 REFERENCES

1. C. Trungpa, *Journey Without Goal: The Tantric Wisdom of the Buddha*, Shambhala Publications, 2010.
2. Sharma H. (2015). Meditation: Process and effects. *Ayu*, *36*(3), 233–237. https://doi.org/10.4103/0974-8520.182756.
3. Heather Buttle, "Measuring a Journey without Goal: Meditation, Spirituality, and Physiology", BioMed Research International, vol. 2015, Article ID 891671, 8 pages, 2015. https://doi.org/10.1155/2015/891671.
4. J. Piet and E. Hougaard, "The effect of mindfulness-based cognitive therapy for prevention of relapse in recurrent major depressive disorder: a systematic review and meta-analysis," *Clinical Psychology Review*, vol. 31, no. 6, pp. 1032–1040, 2011.
5. Murakami H, Nakao T, Matsunaga M, Kasuya Y, Shinoda J, Yamada J, et al. (2012) The Structure of Mindful Brain. PLoS ONE 7(9): e46377. https://doi.org/10.1371/journal.pone.0046377.
6. Fayed N, Lopez del Hoyo Y, Andres E, Serrano-Blanco A, Bellón J, Aguilar K, et al. (2013) Brain Changes in Long-Term Zen Meditators Using Proton Magnetic Resonance Spectroscopy and Diffusion Tensor Imaging: A Controlled Study. PLoS ONE 8(3): e58476. https://doi.org/10.1371/journal.pone.0058476.
7. Santarnecchi E, D'Arista S, Egiziano E, Gardi C, Petrosino R, Vatti G, et al. (2014) Interaction between Neuroanatomical and Psychological Changes after Mindfulness-Based Training. PLoS ONE 9(10): e108359. https://doi.org/10.1371/journal.pone.0108359.
8. Parmentier, F., García-Toro, M., García-Campayo, J., Yañez, A. M., Andrés, P., & Gili, M. (2019). Mindfulness and Symptoms of Depression and Anxiety in the General Population: The Mediating Roles of Worry, Rumination, Reappraisal and Suppression. *Frontiers in psychology*, *10*, 506. https://doi.org/10.3389/fpsyg.2019.00506.
9. Skokou, M., Soubasi, E., &Gourzis, P. (2012). Depression in multiple sclerosis: a review of assessment and treatment approaches in adult and pediatric populations. *ISRN neurology*, *2012*, 427102. https://doi.org/10.5402/2012/427102.
10. Simmons RD, Ponsonby AL, van der Mei IA, Sheridan P. What affects your MS? Responses to an anonymous, Internet-based epidemiological survey. MultScler. 2004 Apr;10(2):202-11. doi: 10.1191/1352458504ms1006oa. PMID: 15124768.

11. Esmonde L, Long AF. Complementary therapy use by persons with multiple sclerosis: benefits and research priorities. Complement Ther Clin Pract. 2008 Aug;14(3):176-84. doi: 10.1016/j.ctcp.2008.03.001. Epub 2008 May 7. PMID: 18640629.
12. Adam B. Levin, Emily J. Hadgkiss, Tracey J. Weiland, Claudia H. Marck, Dania M. van der Meer, Naresh G. Pereira, George A. Jelinek, "Can Meditation Influence Quality of Life, Depression, and Disease Outcome in Multiple Sclerosis? Findings from a Large International Web-Based Study", Behavioural Neurology, vol. 2014, Article ID 916519, 9 pages, 2014. https://doi.org/10.1155/2014/916519.
13. R. Manocha, D. Black, J. Sarris, C. Stough, "A Randomized, Controlled Trial of Meditation for Work Stress, Anxiety and Depressed Mood in Full-Time Workers", Evidence-Based Complementary and Alternative Medicine, vol. 2011, Article ID 960583, 8 pages, 2011. https://doi.org/10.1155/2011/960583.
14. Berry, J. W. in Handbook of multicultural perspectives on stress and coping 287–298 (Spring, 2006).
15. Lopez-Maya E, Olmstead R, Irwin MR (2019) Mindfulness meditation and improvement in depressive symptoms among Spanish- and English speaking adults: A randomized, controlled, comparative efficacy trial. PLoS ONE 14(7): e0219425. https://doi.org/10.1371/journal.pone.0219425.
16. Elder, C., Nidich, S., Moriarty, F., &Nidich, R. (2014). Effect of transcendental meditation on employee stress, depression, and burnout: a randomized controlled study. *The Permanente journal*, *18*(1), 19–23. https://doi.org/10.7812/TPP/13-102.
17. Sheppard WD, II, Staggers FJ, John L. The effects of a stress-management program in a high security government agency. Anxiety Stress Coping. 1997;10(4):341–50. DOI: http://dx.doi.org/10.1080/10615809708249308.
18. Yunesian, M., Aslani, A., Vash, J. H., &Yazdi, A. B. (2008). Effects of Transcendental Meditation on mental health: a before-after study. *Clinical practice and epidemiology in mental health : CP & EMH*, *4*, 25. https://doi.org/10.1186/1745-0179-4-25.
19. Rashi Sharma (2018). To study the effect of Gayatri mantra on Manas Prakruti and stress. International Journal of Engineering Development and Research. Volume 6, Issue 2 | ISSN: 2321-9939.

20. Julie Lynch, Lucia Prihodova, Pádraic J. Dunne, Áine Carroll, Cathal Walsh, Geraldine McMahon, Barry White (2018) Mantra meditation for mental health in the general population: A systematic review,
21. European Journal of Integrative Medicine 23: 101-108,ISSN 1876-3820, https://doi.org/10.1016/j.eujim.2018.09.010.
22. Kumar, S., Nagendra, H., Manjunath, N., Naveen, K., &Telles, S. (2010). Meditation on OM: Relevance from ancient texts and contemporary science. *International journal of yoga*, *3*(1), 2–5. https://doi.org/10.4103/0973-6131.66771.
23. Telles S, Nagarathna R, Nagendra HR. Autonomic changes during "OM" meditation.Indian J PhysiolPharmacol. 1995 Oct; 39(4):418-20.\
24. Kalyani, B. G., Venkatasubramanian, G., Arasappa, R., Rao, N. P., Kalmady, S. V., Behere, R. V., Rao, H., Vasudev, M. K., & Gangadhar, B. N. (2011). Neurohemodynamic correlates of 'OM' chanting: A pilot functional magnetic resonance imaging study. *International journal of yoga*, *4*(1), 3–6. https://doi.org/10.4103/0973-6131.78171.
25. Ira Das ,Himani Anand. Effect of Prayer and "OM" Meditation in Enhancing Galvanic Skin Response. Psychological Thought, 2012, Vol. 5(2), 141–149, doi:10.5964/psyct.v5i2.18.
26. Bhavna P. Harne , Azra A. Tahseena , Anil S. Hiwaleb , R. S. Dhekekar. Survey on Om Meditation: Its Effects on the Human Body and Om Meditation as a Tool for Stress Management. Psychological Thought, 2019, Vol. 12(1), 1–11, https://doi.org/10.5964/psyct.v12i1.275.
27. Kjellgren, A., Bood, S. A., Axelsson, K., Norlander, T., &Saatcioglu, F. (2007). Wellness through a comprehensive yogic breathing program - a controlled pilot trial. *BMC complementary and alternative medicine*, *7*, 43. https://doi.org/10.1186/1472-6882-7-43
28. Seppälä, E. M., Nitschke, J. B., Tudorascu, D. L., Hayes, A., Goldstein, M. R., Nguyen, D. T., Perlman, D., & Davidson, R. J. (2014). Breathing-based meditation decreases posttraumatic stress disorder symptoms in U.S. military veterans: a randomized controlled longitudinal study. *Journal of traumatic stress*, *27*(4), 397–405. https://doi.org/10.1002/jts.21936
29. Zope, S. A., &Zope, R. A. (2013). Sudarshan kriya yoga: Breathing for health. *International journal of yoga*, *6*(1), 4–10. https://doi.org/10.4103/0973-6131.105935

30. Vinod D. Deshmukh, "Neuroscience of Meditation", The Scientific World Journal, vol. 6, Article ID 740623, 15 pages, 2006. https://doi.org/10.1100/tsw.2006.353.

31. Tran, U. S., Cebolla, A., Glück, T. M., Soler, J., Garcia-Campayo, J., & von Moy, T. (2014). The serenity of the meditating mind: a cross-cultural psychometric study on a two-factor higher order structure of mindfulness, its effects, and mechanisms related to mental health among experienced meditators. *PloS one*, *9*(10), e110192. https://doi.org/10.1371/journal.pone.0110192

32. Vieten C, Wahbeh H, Cahn BR, MacLean K, Estrada M, Mills P, et al. (2018) Future directions in meditation research: Recommendations for expanding the field of contemplative science. PLoS ONE 13(11): e0205740. https://doi.org/10.1371/journal.pone.0205740.

INDEX

N

O

P